Palgrave Studies in the History of Science and Technology

Designed to bridge the gap between the history of science and the history of technology, this series publishes the best new work by promising and accomplished authors in both areas. In particular, it offers historical perspectives on issues of current and ongoing concern, provides international and global perspectives on scientific issues, and encourages productive communication between historians and practicing scientists.

More information about this series at
http://www.palgrave.com/gp/series/14581

Zoltán Boldizsár Simon

The Epochal Event

Transformations in the Entangled Human, Technological, and Natural Worlds

palgrave
macmillan

Zoltán Boldizsár Simon
Bielefeld University
Bielefeld, Germany

Palgrave Studies in the History of Science and Technology
ISBN 978-3-030-47804-9 ISBN 978-3-030-47805-6 (eBook)
https://doi.org/10.1007/978-3-030-47805-6

Cover illustration: Julia Filimonova

This Palgrave Pivot imprint is published by the registered company Springer Nature Switzerland AG.
The registered company address is: Gewerbestrasse 11, 6330 Cham, Switzerland

Preface: On Connective Concepts

Human activity confronts us today with our cognitive limits. The recognition that human agency achieved a planetary character, that is, the recognition that our activity became a natural force that transforms the condition of the Earth system, is at the edge of our understanding. Facing the sheer scale and the immense consequences of human actions and capacities does not come easy to most of us. Little wonder that the new situation triggers contradictory responses. Only in the climate context, responses vary from climate change denial to apocalyptic fatalism, from lifestyle choices of reducing carbon footprints to large-scale plans of geoengineering, and from technology blaming to techno-saviorism.

In a cacophony of voices, scholarly responses represent only a fragment of possibilities. Even if the natural sciences brought the issue to a broader awareness in the first place, their voice becomes one among the many when it comes to responding to the situation. Together with humanities and social scientific responses, they constitute a larger family of responses with a specific socio-cultural function to fulfill: scholarly responses do not simply confront us with the limits of understanding; they also try to overcome the very limits they identify. In other words, scholarly responses aim at gaining an understanding of that which seems to defy understanding—especially in situations we experience as crisis.

This book attempts to sketch a specific type of scholarly response. It aims at developing an understanding of our recent anthropogenic planetary crisis by creating concepts through which we conceive of the world and ourselves. So far this is nothing surprising. Many would agree that theoretically oriented work in what we still call today the humanities and the social sciences is, to a large extent, conceptual work.

The specificity of this book's approach lies in the intention to encourage developing concepts of a peculiar kind. Let's call them *connective concepts.* As our current crisis consists of the collision of human and the natural worlds through advanced technologies, the adequate concepts through which we apprehend ourselves and the world must be ones that somehow capture the collision of worlds or reflect on that collision. Such concepts are connective inasmuch as they have the potential to link, bridge, and connect knowledge formations that were originally developed to study the human and the natural worlds separately. In order to live up to their potential, they need to travel.

As Mieke Bal pointed out two decades ago, concepts "are not fixed"; "they travel – between disciplines, between individual scholars, between historical periods, and between geographically dispersed academic communities."[1] The different forms of travel Bal indicates are supposed to testify the flexibility of concepts. How do connective concepts relate to Bal's traveling concepts and their flexibility? On the one hand, connective concepts are flexible to the extent that they, too, travel across disciplines and their meanings are constantly being renegotiated. On the other hand, connective concepts also gesture toward the opposite end by seeking points of connection and shared meanings through potentially the entirety of the scholarly landscape, thereby enabling a broader exchange on the collision of worlds. As compared to traveling concepts, connective ones have an extra job to do. Whereas Bal's notion of traveling concepts is primarily a reflection on the work of the humanities (even if she does not exclude science from the travels), connective concepts, by definition, may link not only certain disciplines, but, in the most desirable although seldom cases, also the work of the humanities and the work of the sciences.

At times when we both unintentionally act into and deliberately manipulate and engineer what has previously been thought of as the order of nature, concepts that merely travel are less useful than concepts that also connect. To stand a chance to comprehend the big picture, our partial knowledges must be effectively bridged by certain concepts. In fact, even the question of how exactly we understand the collision of the human and the natural worlds—whether it makes up one world, whether we keep thinking about them as distinguishable (and if yes, the way in which we do exactly), or whether we somehow manage to think of it as one world with analytically distinguishable constituents—will likely be decided by the concepts we create to capture their entanglement.

The collision of worlds is already enhanced into a large-scale connective concept, even a master concept, if you like: the Anthropocene. Yet, as of now, the Anthropocene is also a contested concept in a transdisciplinary debate, invested with several potentially conflicting meanings. The chapters of this book will devote a great deal of attention to exploring the perplexities and conflicts in debating the concept. What I would like to do here is to indicate four points about connective concepts by taking a starting point in the notion of the Anthropocene. Let me introduce the points in the shape of four brief theses.

1. *Connective concepts entail a cluster of conceptual innovations.* Game-changing conceptual innovations hardly concern stand-alone concepts. Rather, they involve a larger cluster or web of interrelated notions, some of which are newly coined, while others are old ones invested with new meanings. In the case of the Anthropocene, its fate as a master concept is tied to the development of further concepts that either reinforce or challenge its plausibility. Hence the fact that the Anthropocene already comes within a rapidly expanding web of concepts that intend to capture the current "planetary" and "anthropogenic" crisis.
2. *Connective concepts require mutual knowledge transfer.* Concepts typically emerge locally and become connective ones over the course of multiple interactions. The Anthropocene as a connective (master) concept is, for instance, the creation of Earth system science, just as well as most of its surrounding concepts. The humanities and the social sciences have begun to rapidly adopt these concepts among their conceptual tools, renegotiate their meanings, and develop their own concepts that either function as new surrounding and related concepts to the scientific ones or as conceptual alternatives. This is necessary, and conceptual work in this direction is unlikely to halt anytime soon. As seen from the other side, however, the conceptual innovations of the humanities and social sciences struggle to find their way to the scientific vocabulary to a comparable extent. Mildly put, at the present stage it is only in a limited sense that we can meaningfully talk about a mutual knowledge transfer in making sense of the collision of worlds. The reasons are manifold, and some of them will be touched upon in more details in the chapters that

follow. What needs to be noted here is only the fact that the ideal of mutuality in knowledge transfer can easily fall short in practice. Yet, the lack of full mutuality does not mean that partial mutuality is of no value. Under the condition of contested meanings—and concepts discussed across the disciplinary landscape are extremely prone to such contestation of meanings—partial mutuality is already a success.

3. *With respect to their reach, first-order and second-order connective concepts can be distinguished.* Given the distance from the ideal situation of achieving a high degree of mutuality in knowledge transfer, it makes sense to distinguish between two levels in which connective concepts may work. On the first level, one can find those *first-order connective concepts* that resonate with both the natural and life sciences on the one hand and the human and social sciences on the other. Such concepts are, I believe, what we need, what we are short of, what we bitterly contest even when we have them, and what may be seldom and extremely difficult to intentionally develop. As an example of a truly transdisciplinary reach, one can think of not only the Anthropocene but also the notion of anthropocentrism, fiercely debated in the humanities and the social sciences as well as in the natural and life sciences.[2] *Second-order connective concepts* have a narrower reach. They work, respectively, within the confines of the natural and life sciences on the one hand and the human and social sciences on the other. Arguably, much of the conceptual innovations that emerge locally in disciplines of the human and social sciences acquire such reach at best. They may resonate broadly enough within humanities and social scientific scholarship, but they hardly find their way to the scientific vocabulary. This is especially true of concepts put forward as alternatives to the Anthropocene, such as the Capitalocene or the Chthulucene.[3]
4. *Connective concepts are recognized as such only after the fact and by being used as connective ones.* The above categorization of connective concepts does not entail that each individual concept must live up to the fullest possible reach within one of the categories. It means only that connective concepts are typically made use of within the respective scopes associated with them. As the phrasing of the previous sentence already indicates, concepts are defined as connective by virtue of being used as such. Concepts may or may not be developed with the explicit intention to function as connective ones, but

> intention is not what decides their connective character. Nor is the decisive factor anything like an inherent feature of the concepts themselves. We are the ones who invest concepts with a connective character, and we do that only after the fact, only after the formation of the concepts, by actually putting them to use across the disciplinary landscape as conceptual bridges.

This book is an effort to develop such a connective concept: the epochal event. The concept reflects the emerging societal experience of time that epochal changes are taking place all around us. Against the backdrop of the technology-driven collision of the human and the natural worlds, it attempts to capture the transformative character of threshold events that trigger previously unimaginable epochal transformations.

Bielefeld, Germany
Leiden, The Netherlands

Zoltán B. Simon

Notes

1. Mieke Bal, *Travelling Concepts in the Humanities: A Rough Guide* (Toronto: University of Toronto Press, 2002), 24.
2. In the larger debate on anthropocentrism, the humanities are typically advocating an anti-anthropocentric stance. Chapter 4 engages with the question of anthropocentrism in more details.
3. More on these alternatives in Chapter 3.

Bibliography

Bal, Mieke. *Travelling Concepts in the Humanities: A Rough Guide*. Toronto: University of Toronto Press, 2002.

Acknowledgments

Not many friends knew that I was working on this book. The reason for this is very simple: neither did I. Around the end of 2018, I started to work on an article, which, over time, became a long article. Then it became a very long one—so long, after a while, that at some point I began to entertain the idea of the book format. And now here we are.

Writing this book coincided with the blossoming of my collaboration with Marek Tamm on various related projects. I am indebted to Marek and our discussions of many themes and ideas that feature in the coming pages. The final shape of the text greatly benefited also from the comments of the reviewers, pointing at potential pitfalls and indicating ways to strengthen arguments. Very many thanks to Julia Filimonova for drawing the brilliant picture for the cover and for Megan Laddusaw at Palgrave for managing this project through all its stages, from proposal to the published book.

Contents

CHAPTER 1

Prelude to a New Epochality

Abstract An emerging societal experience of time conceives of changes in epochal terms. Epochal claims increasingly dominate the ways in which we think about transformations both of the world and of knowledge production about the world. To a certain extent, the latter seems self-evident: insofar as we think that epochal changes are around us, it seems reasonable to claim that we need epochal changes in our modes of thinking to be able to come to terms with the vastness of the changes themselves. In briefly introducing the emergence of a new epochality, that is, a new kind of epochal thinking, the chapter touches upon some of the core themes of the book: the perception that human capacities are reaching an unprecedented scale; the entanglement of the human, technological, and natural worlds; the shift toward thinking in planetary terms; and the necessity to develop a knowledge regime attuned to the new conditions. Finally, the chapter ends with an outline of the book.

Keywords Epochal change • Epochality • Experience of time • Human • Technology • Nature • Planetary • Transdisciplinarity

Z. B. Simon, *The Epochal Event*, Palgrave Studies in the History of Science and Technology,
https://doi.org/10.1007/978-3-030-47805-6_1

The New Epochality

We are living in an age unlike anything the world has ever seen. This is not a claim I wish to advance but a societal experience of time that I think we need to understand. The new experience of time is historical in a specific way: it conceives of changes in epochal terms. Not in the most familiar sense though, not in the sense of historical epochs as we know them. Epochal thought is emerging today in a way that was simply inconceivable before human capacities came to be perceived in terms of a planetary-scale agency that brings about transformations in the condition of the Earth viewed as an integrated system.

Such human agency lurks behind an unprecedented biodiversity loss, the alteration of the climatic conditions of the planet, and the potential to manipulate and create both non-organic and organic life-forms through advanced technologies (artificial intelligence, digital life-forms, genome editing, cloning, and so on). All this, either separately or together, is expected to launch epochal changes on a planetary scale, entangle the human and the natural worlds through advanced technology, and kick off transformations with potentially unpredictable consequences that the human mind is not even equipped to grasp. What we can grasp, however, is the societal experience of the epochal transformations that we are likely to bring about. This book is an effort to come to terms with a potentially new kind of epochal thinking—rooted in the perception of anthropogenic planetary changes—that endows even certain towering events with epochal attributes.

But let's not get too ahead of ourselves and begin with the most fundamental experiences. There is arguably a growing sense that epochal changes are taking place around us in practically all domains of life. Following the invention of nuclear weaponry and the experiences of the Second World War, we have been said to enter the atomic age. Then came the space age, and subsequently the information age, while we have been undergoing a digital revolution, which is now already old news in light of the fourth industrial revolution that brings together the digital of the third industrial revolution with physical systems. Perhaps some would argue that it is a feature of postmodernity (understood in epochal terms as a societal or socio-cultural condition) that industrial revolutions appear to come lately in sequels, much like Hollywood blockbusters. Yet there is a chance that these claims would fall on deaf ears, as today postmodernity itself as an era is gone, and the intellectual period when postmodernism was a prominent and fashionable mode of thinking (not be confused with postmodernity) is also passé. It seems we have now entered the Anthropocene, with our eyes scanning the horizon for the coming posthumanity.

Anyone can add their favorite epochal claims to the ones listed here. I also add one by writing this book and arguing that coming to terms with our new epochality, that is, with our *age of the epochal*, so to speak, requires some sort of an intellectual sea change. Besides, this claim will fit quite smoothly with a tendency in present-day scholarship that reproduces the broader societal experience of time on its own scale. Given that scholarly endeavors take place within the broader societal space, it is not much of a surprise that the wider societal sentiment of the epochal boils down to epochal claims concerning knowledge production. In regard to knowledge production, we have gotten used to framing intellectual endeavors as outlines of colossal challenges that demand responses which entail tremendous changes in our modes of thinking.

Again, my intention is not to point fingers. I have been exercising such scholarship before, and this book is no exception. I only want to raise awareness of the fact that this is what we are doing, oftentimes without knowing it. We are of course very well aware of the customary academic turns that typically argue for refocusing attention and aim at *re*thinking, *re*figuring, and *re*inventing that which already exists. Epochal claims, however, aim at a much higher goal; they demand changes on the level of wholesale knowledge formations. And, to a certain extent, this scholarly reproduction of the wider societal experience may even be self-evident. For insofar as we think that epochal changes are around us, it seems reasonable to claim that we need epochal changes in our modes of thinking to understand them.

Here are a few examples. To begin with, consider Donna Haraway's discussion of "the almost incomprehensible increases in human numbers." More concretely, Haraway refers to "a 9 billion increase of human beings over 150 years, to a level of 11 billion by 2100 if we are lucky," which "cannot be explained away by blaming Capitalism or any other word starting with a capital letter." Instead, the new situation points onward to the need "to think together anew across differences of historical position and of kinds of knowledge and expertise."[1] Haraway's implied take on the necessity of new knowledge formations is nevertheless less explicit than that of Elizabeth Ermarth (in a context that does not have much to do with planetary-scale changes). In her book *History in the Discursive Condition*, published less than a decade ago to synthetize her work from the two previous decades, Ermarth puts forward the argument that the "modern condition" has been replaced by a "discursive condition." She argues that "the departure from modernity signals a tectonic change radical enough to suggest that even the tools of thought must change if we are to keep up with ourselves in any vital or creative way."[2]

Less than ten years later, however, there are not many left who still think that we are in anything like a "discursive condition." Yet, perhaps even more of us think that we are witnessing "a tectonic change radical enough" to rewrite previous knowledges, even though the reasons for thinking so may be other than Ermarth's. It makes sense then to return to potentially more future proof final examples, ones that reflect the planetary concerns integral to the coming chapters. One of the first things to mention in this context is the human-induced mass extinction of species. In their 1996 book *The Sixth Extinction*, Richard Leakey and Roger Lewin argue that the recognition of a sixth mass extinction in Earth's history demands an epochal change in the constitution of knowledge. They claim that "we are in the midst of a seismic shift in thinking about the nature of ourselves and the world we live in. It is no hyperbole to describe the magnitude of the shift as an intellectual revolution."[3] Then, in the context of anthropogenic climate change—and in a more moderate and less alarmist tone—Tracey Skillington even better captures the way in which epochal claims about the world and knowledge production on the world hang together. In her recent book *Climate Justice and Human Rights*, Skillington writes that "the anticipation of grave environmental catastrophes fundamentally alters ways of being in and thinking about this world."[4]

Again, anyone can add their favorite epochal claims concerning the scholarly world of knowledge production. But before all this begins to look too much like a parody (with respect to both scholarship in the age of epochal thinking and the age itself as some sort of a meta-epoch), I must make it perfectly clear that the phenomenon is genuinely serious. We are so invested in advocating our particular epochal claims that we fail to see the big picture of what exactly we are doing. More importantly, in failing to see how we advance epochal thought in the scholarly world, we also fail to understand that which we somewhat automatically and habitually reproduce: the larger pattern of thought, the broader societal mentality that conceives of the world not merely in terms of accelerating changes (as recent social theories hold) but in terms of non-continuous epochal transformations.[5]

To avoid misunderstandings, there is nothing wrong with the scholarly reproduction of the epochal. The relative unawareness about it is a bit more problematic though. I intend to remedy this situation to the extent of bringing to the fore a category of thought already implicit in our age of the epochal. The ultimate aim of this book is to account for the emergence of events which are perceived as having the capacity to kick off epochal changes. I call them *epochal events*. Achieving this aim implies two supplementary undertakings. First, I need to outline the general conditions under which such epochal events can emerge. This means that—on a more general level—I have to account for the peculiarities of the new kind of epochal thinking that permeates societal and scholarly imagination since the middle of the twentieth century. Second, being subjected to the very thinking that I examine, I will end up advancing scholarly claims concerning knowledge production—claims that must also come out as epochal, very much in the fashion of the aforementioned examples, although hopefully more self-consciously.[6]

The Structure of the Book

To briefly foreshadow the gist of the coming pages, its epochal claims concern the necessity of venturing into a transdisciplinary knowledge regime designed to study the new constellation of the entanglement of the human, the technological, and the natural worlds.

Through developed technologies, human activity is merged today with the natural world to the extent that planetary-scale transformations emerge within an entanglement of worlds. The entanglement—and the growing awareness of the human element being effective in planetary changes—provides the context in which epochal thought gains new life. Yet we struggle to understand even the stakes of the entanglement if we remain constrained by our disciplinary knowledge formations that have been established on the premise of studying the human and the natural worlds separately. If these worlds appear to us as intertwined and interconnected today, then we need to develop the matching knowledge regime that brings together and connects humanities and social scientific scholarship with the work of the natural and life sciences. As many approaches with which the coming pages enter into a discussion already argue along similar lines, the question is not that of whether it is necessary to gesture toward a transdisciplinary direction in knowledge production. The question is rather how exactly to do so.

This book intends to develop an answer to the above question by explicating the epochal event as a *connective concept* with a transdisciplinary appeal. Following this introduction, Chapter 2 sets the stage by arguing that at times when a new kind of epochal imagination pervades both public and scientific discourses, there is an appeal to history. The appeal turns out to be a concern not of history as disciplinary knowledge but of history as a mode of thinking. Besides, it entails a demand for history to renegotiate itself, that is, to renegotiate the narrowly understood disciplinary epistemology of historical studies and whatever we mean by history today. All in all, amidst epochal changes in the entangled human-technological-natural world, an appeal to history must be an appeal to develop the kind of historical thinking that is able to make sense of transformations in a more-than-human world.

The task of Chapter 3 is to explore that which evokes a potentially renewed historical thinking: the entanglement of the world of human affairs, the technological domain, and the natural world. Focusing on the Anthropocene debate, the chapter pays special attention to the question of how technology qualifies as a systemic element in the overall entanglement. It also features a sustained argument for the necessity of developing concepts that enable us to come to terms with the entanglement—concepts through which we can be able to see that there is an entanglement in the first place. Such concepts need to be connective concepts, that is, concepts that find resonance across the disciplinary landscape and gesture toward the possibility of developing the knowledge formations adequate for studying the entanglement of the human, technological, and natural worlds.

Chapter 4 outlines the emergence of the new epochality in post-Second World War societies. It argues that a new kind of epochal thought is being constituted since the middle of the last century, since humanity's self-authored existential catastrophe—within the context of the entangled human-technological-natural world—became thinkable. The chapter introduces three instances of the new epochal: first, the ongoing human-induced sixth mass extinction event mentioned earlier[7]; second, the potential transgression of planetary boundaries, in which the notion of "planetary boundaries" attempts to identify a safe operating space for humanity within the frontiers that mark the conditions of the planet that, to our present knowledge, support human life[8]; and third, the prospect of a technological singularity, the potential event of creating greater-than-human general intelligence, likely entailing an intelligence explosion in creating increasingly greater intelligences in an accelerating pace.[9] In relation to the three instances of the new epochal, the chapter also ventures into a discussion of several key themes, ranging from the question of anthropocentrism to the way in which the future appears as a threat. The central significance of Chapter 4 lies with the fact that it is within the epochal thought emerging in the shadow of anthropogenic catastrophe that certain events achieve an epochal character.

Whereas the chapters mentioned so far are devoted to big questions concerning the larger context and the condition of possibility of epochal events, the subsequent two are explicitly devoted to the conceptualization of the epochal event as a category of a renewed historical thought. Chapter 5 begins the conceptual work by turning to the closest available category to which the epochal event can be measured: the historical event. In mapping existing conceptualizations of the historical event, the chapter focuses on the relationship between event and novelty and on the transformative potential of events. Three distinct notions of the historical event receive special attention: Rolf Gruner's analysis of the historical event as an exercise in analytic philosophy; William Sewell's thoughts on historical events that transform social structures; and Hayden White's "modernist" event.

Chapter 6 situates these notions of the historical event with the new instances of the epochal introduced in Chapter 4. Over the course of building a contrast, the epochal event proves to be a distinct category—that is, not merely a historical event of a larger scale and a bigger scope—endowed with the following characteristics: first, epochal events are hyper-historical, meaning that the transformations they induce open up a "new reality"; second, they are perceived as having a potential to introduce thereby a radical temporal break interpreted in terms of temporal incommensurability; third, they exceed the confines of human experience. To condense all this (and more) into a digestible takeaway message, the chapter attempts to develop a definition of the epochal event.

Finally, in place of a conclusion, the coda returns to the big picture with a dual objective. On the one hand, it reviews related efforts that, in one way or another, grapple with conceptualizing a world of epochal transformations in a more-than-human world. On the other hand, it muses over the possibility of learning to inhabit such a world. As the final contention of the book, the coda suggests that inhabiting a world of epochal transformations might mean nothing other than learning to navigate carefully among oftentimes contradictory imperatives.

Notes

1. Donna J. Haraway, *Staying with the Trouble: Making Kin in the Chthulucene* (Durham: Duke University Press, 2016), 6–7.
2. Elizabeth Deeds Ermarth, *History in the Discursive Condition: Reconsidering the Tools of Thought* (London and New York: Routledge, 2011), xii.
3. Richard Leakey and Roger Lewin, *The Sixth Extinction: Patterns of Life and the Future of Humankind* (New York: Anchor Books, 1996), 223.
4. Tracey Skillington, *Climate Justice and Human Rights* (New York: Palgrave, 2017), 2. The Anthropocene debate across disciplines—introduced properly in the next chapters—features countless claims concerning the necessity of epochal changes in knowledge production. So does this book.
5. For the most influential sociological theory of acceleration, see Hartmut Rosa, *Social Acceleration: A New Theory of Modernity*, trans. Jonathan Trejo-Mathys (New York: Columbia University Press, 2013). For a rival sociological interpretation—sensing that what is stake is more than acceleration—see Ulrich Beck, *The Metamorphosis of the World* (Cambridge: Polity, 2016). The coda will return to Beck's theory in the context of discussing endeavors that share the main imperative of this book and grapple with providing a conceptual understanding of an age of epochal transformations. For my more detailed discussion of Rosa's theory in the terms outlined above, see Zoltán Boldizsár Simon, *History in Times of Unprecedented Change: A Theory for the 21st Century* (London: Bloomsbury, 2019), 181–183.
6. Whether the growing societal and scholarly sense of witnessing epochal transformations can be confined to Western societies, to the experiences of the rich, or generally speaking to the experiences of those who can afford to experience the world in such terms is of course an open question. Globalization makes it difficult to argue for confining claims concerning societal tendencies to the Western world, let alone, to certain social groups within the Western world. At the same time, the categories of the humanities and the social sciences that point to various inequalities within social worlds make it difficult to advance general claims concerning large-scale societal tendencies, not to mention extending them over the globe. There is no easy way out of this trap. I am inclined to leave the issue unresolved by making peace with the fact that even if I attempted a resolution it would be bitterly challenged and everyone would situate the outreach of this epochal sentiment as they themselves would consider it adequate.
7. See also Stuart J. Pimm and Thomas M. Brooks, "The Sixth Extinction: How Large, Where, and When?," in *Nature and Human Society: The Quest for a Sustainable World*, eds. Peter H. Raven and Tania Williams (Washington, DC: National Academy Press, 2000), 46–62; Elizabeth Kolbert, *The Sixth Extinction: An Unnatural History* (London: Bloomsbury, 2014).

8. Johan Rockström et al., "Planetary Boundaries: Exploring the Safe Operating Space for Humanity," *Ecology and Society* 14, no. 2 (2009): art. 32; Will Steffen et al., "Planetary Boundaries: Guiding Human Development on a Changing Planet," *Science* 347, no. 6223 (2015): 1259855.
9. The term was put into wider circulation by Vernor Vinge, "The Coming Technological Singularity: How to Survive in the Post-Human Era," in *Vision-21: Interdisciplinary Science and Engineering in the Era of Cyberspace*, Proceedings of a symposium cosponsored by the NASA Lewis Research Center and the Ohio Aerospace Institute, Westlake, Ohio, March 30–31 (1993), 12–13. For a broader introduction to the theme, see Murray Shanahan, *The Technological Singularity* (Cambridge, MA: MIT Press, 2015).

Bibliography

Beck, Ulrich. *The Metamorphosis of the World*. Cambridge: Polity, 2016.

Ermarth, Elizabeth Deeds. *History in the Discursive Condition: Reconsidering the Tools of Thought*. London and New York: Routledge, 2011.

Haraway, Donna J. *Staying with the Trouble: Making Kin in the Chthulucene*. Durham: Duke University Press, 2016.

Kolbert, Elizabeth. *The Sixth Extinction: An Unnatural History*. London: Bloomsbury, 2014.

Leakey, Richard, and Roger Lewin. *The Sixth Extinction: Patterns of Life and the Future of Humankind*. New York: Anchor Books, 1996.

Pimm, Stuart J., and Thomas M. Brooks. "The Sixth Extinction: How Large, Where, and When?," In *Nature and Human Society: The Quest for a Sustainable World*, edited by Peter H. Raven and Tania Williams, 46–62. Washington, DC: National Academy Press, 2000.

Rockström, Johan et al. "Planetary Boundaries: Exploring the Safe Operating Space for Humanity." *Ecology and Society* 14, no. 2 (2009): art. 32.

Rosa, Hartmut. *Social Acceleration: A New Theory of Modernity*. Translated by Jonathan Trejo-Mathys, New York: Columbia University Press, 2013.

Shanahan, Murray. *The Technological Singularity*. Cambridge, MA: MIT Press, 2015.

Simon, Zoltán Boldizsár. *History in Times of Unprecedented Change: A Theory for the 21st Century*. London: Bloomsbury, 2019.

Skillington, Tracey. *Climate Justice and Human Rights*. New York: Palgrave, 2017.

Steffen, Will et al. "Planetary Boundaries: Guiding Human Development on a Changing Planet." *Science* 347, Issue 6223 (2015): 1259855.

Vinge, Vernor. "The Coming Technological Singularity: How to Survive in the Post-Human Era." In *Vision-21: Interdisciplinary Science and Engineering in the Era of Cyberspace*, 11–22. Proceedings of a Symposium Cosponsored by the NASA Lewis Research Center and the Ohio Aerospace Institute, Westlake, Ohio, 30–31 March 1993.

CHAPTER 2

A Perplexing Appeal to History

Abstract The emergence of a new epochality entails an appeal to history, but the epistemological foundations of the discipline of history do not enable adequate responses. Whereas twenty-first-century thinking conceives of epochal changes in the context of human activity being inextricably intertwined with nature through technology, disciplinary history has been institutionalized in the nineteenth century on the premise of studying a specifically human world. The chapter resolves this question by arguing that the appeal to history concerns a mode of thinking that cannot be reduced to a narrowly understood disciplinary epistemology. In Western modernity, historical thinking conquered not only the human but (through the idea of evolution in biology and gradual Earth history in geology) also the natural world. The modern distribution of work in writing human and natural histories, respectively, by history and the sciences, however, can hardly be maintained. Accordingly, the appeal to history as a mode of thinking is accompanied today by a demand to develop a new kind of historical thinking attuned to studying a more-than-human world (more-than-human as seen from disciplinary history).

Keywords Historical thinking • Historicization • Disciplinary confines • Epistemology • Anthropocene • More-than-human

Z. B. Simon, *The Epochal Event*, Palgrave Studies in the History of Science and Technology,
https://doi.org/10.1007/978-3-030-47805-6_2

Setting the Stakes

Historical events are, so to speak, already history. The pivotal role they played in modern historical thinking is now occupied by another kind of event. Historical events are history at least in two respects: first, in the sense of being a fundamental category of historical thought as applied to the ever-changing realm of human affairs, and second, in the sense that we tend to think of historical events only retrospectively, only as events that already took place. In both respects, the notion of the historical event seems far too narrow to adequately capture the ways in which twenty-first-century thinking conceives of events that signal momentous change.

The events that fascinate present-day imagination are what I call *epochal events*. By this, I mean events conceived of as markers of epochal transformations that escape the confines of the course of human affairs and apply prospectively just as well as retrospectively. In the coming chapters, I attempt to mold the notion of the epochal event into a conceptual shape and explicate it as an emerging category of a new kind of historical thinking. For now, the question that must be answered goes as follows: what could it possibly mean that the scope of transformative events extends beyond the human world and even into the future?

To begin with, the aspect of surpassing the scale and scope of the human world does not mean that epochal transformations no longer concern the human world whatsoever. It rather means that the primary (but not exclusive) context of the epochal event is the human world as entangled with the natural world through technology. The entanglement of the human, the technological, and the natural worlds as the primary context of the epochal event is not merely an extension of the human world over the natural or vice versa. It represents a qualitative shift in the way we frame our thinking and look at the world and ourselves. Then, as to the aspect that epochal events can be recognized not only retrospectively but also prospectively, it does not entail any claim about knowing the future the very same way as one can claim to have knowledge of the past.[1] It merely points to the way in which epochal events—in the context of a human-technology-nature entanglement—are expected to take place in the future due to runaway technological changes and anthropogenic transformations in the Earth system.

The prospects are largely familiar by now. On a smaller scale, we are already sadly accustomed to news about biodiversity loss or massive plastic waste in the oceans and in the stomachs of seabirds. Human-induced climate change, its present effects, and its dire future consequences gain broader and broader publicity day by day. A new vocabulary consisting of phrases such as "carbon footprint" is quickly taking root (at least among those who can afford to care), just as well as a general awareness that failing to reduce carbon emissions is likely to lead to undesired and even catastrophic changes both in the condition of the planet and in human societies.

Anthropogenic transformations in nature are not the only prospects to receive growing publicity these days. The perils of artificial intelligence—both in military or everyday use—and the potential of AI technologies to turn democratic ways of life into authoritarian regimes already pertain the wider societal consciousness to one extent or another. Human enhancement technologies are becoming increasingly applied in the medical domain, and even those who somehow miss public debates on the ethics of human enhancement are most probably familiar with the even stronger prospect of a transhuman and more-than-human future due to cinematic and online streaming experiences. Some technological prospects may look more plausible than others, while the potential changes triggered by technology look desirable and apocalyptic at the same time. But all differences aside concerning the desirability of transformations, hardly anyone doubts that advanced technologies bring about spectacular changes.

On a larger scale, however, and on the conceptual level, we still need to come to terms with the big picture of ongoing planetary-scale changes. At stake is not simply the aggregate of all particular changes in each domain separately. What we need to grasp is the nature and the specific character of transformations in the human-technology-nature matrix, which is precisely what the epochal event intends to achieve as a conceptual category. Rendering the concept intelligible and feasible is, however, a long journey. This chapter sets out on the journey by posing the following question: is there any discipline in our current knowledge regime that could claim expertise in mapping a world of self-made epochal transformations and comprehend it as an overall condition framing our lives?

The Confines of Disciplinary Codes

History seems like a self-evident candidate. As is well known, exploring and understanding change in the world of human affairs has been the expertise of the institutionalized discipline of history in Western modernity. As French historian Marc Bloch famously claimed in the middle of the last century in *The Historian's Craft*, the discipline of history is concerned with investigating and understanding human beings in time.[2] A few decades later, Ernst Breisach rather naturally echoed this definition in his seminal study on the history of historiography, claiming that "history deals with human life as it 'flows' through time,"[3] meaning a tiptoeing between accounting for both change and continuity in the human world. Or, to have a more recent example that also reflects current trends, in her book on the constitution of the discipline, Sarah Maza writes that "most historians would probably agree that their task is twofold: to explain the unfolding of change in the past, and to make the people and their places of the time come alive for their readers."[4]

On this basis it seems plausible to claim that, inasmuch as the question of change over time is at stake and inasmuch as the human world is involved in recent epochal imaginaries, they entail an appeal to history. Yet, inasmuch as change concerns a more-than-human world in today's epochal thought, disciplinary understandings of history become less useful. Due to the emphasis on the human in disciplinary epistemology, escaping the confines of the human world entails that we also need to escape the confines of the expertise of historical studies. And the case remains the same even when bracketing the primacy of change over time and trying to look for conceptions of historical work other than that of Bloch, Breisach, and Maza.

In the first half of the last century, Robin G. Collingwood sharply opposed the idea that historical knowledge is concerned with change over time. The alternative view he offered was that history is the knowledge of human thoughts and the human mind (studied through the manifestations of human thoughts in events and actions).[5] In that, Collingwood was in deep agreement with Giambattista Vico's foundational insight from the eighteenth century, according to which humans can have knowledge of that which is made by humans, and, consequently, history (as the course of affairs) can be known precisely because it is made by humans and is a distinctly human history.[6] This view of historical knowledge, needless to say, separates knowledge of the human world and knowledge of the natural world perhaps even more sharply than the views introduced earlier. Consequently, it may be even less promising for efforts that try to understand transformations in the entangled human-technological-natural world.

In arguing that recent challenges posed by anthropogenic climate change mark the collapse of the modern distinction between natural history and human history—and in briefly reviewing the history of the distinction with a focus on the ideas of Vico, Croce, and Collingwood—Dipesh Chakrabarty came to the same conclusion a decade ago.[7] At its theoretical and epistemological core, the discipline of history is simply designed to understand the ever-changing human world. Despite the rather obvious appeal to history in the recent rise of epochal thinking, the theoretical foundations of historical studies do not seem to allow a sufficient engagement with the task. History, in the shape we came to know it in Western modernity, appears to be part of the very challenge to which we seek to respond.

But if not history, then which discipline is suited for making sense of transformations in the entangled human-technological-natural world? Well, it is very likely that none of the disciplines as we know them. All disciplines find themselves today in the same situation as historical studies do. The changes in the world of human affairs, the natural world, and the technological domain seem to be intertwined in ways so complex that neither the humanities and the social sciences nor the STEM disciplines and the life sciences are able to comprehend them by relying solely on their own methods and categories. The respective analytical tools of the modern disciplines have been created to enable them to study some of these worlds and domains, but they are simply not designed for making sense of the entanglements of these worlds and domains or for understanding anything like an overall constellation of their entanglement.

Yet there is the undeniable appeal to history in the recent rise of epochal thought. It forces me to have a closer look at the question, even at the cost of disregarding respectable foundational theories of disciplinary historical scholarship.

The Modern Project of Historicizing Nature and the Human World Alike

Having this closer look means, somewhat paradoxically, approaching "history" in a broader sense. As a general historical sensibility or as a general historical mode of thinking—as an either unreflective or very much conscious mental operation of making sense of individual occurrences within a larger configuration or scheme of change over time that we tend to invest with meaning—history has not been confined to professionalized historical studies in the modern world. What this means is that even if the overall constellation of the entangled human, technological, and natural worlds is no expertise of history as an *institutionalized discipline*, changes in the entangled words may nevertheless be approached by *a mode of thinking* that is not anchored in disciplinary protocols.

Being "historical" has not been a quality attributed solely to the world of human affairs in Western modernity. The natural and the human worlds have acquired their respective "historical" character and have been historicized thoroughly side by side over the last two centuries or so. True enough, the work of historicization has been carried out typically by disciplines with restricted scope concerning both worlds. But once a sense of historicity has been developed as a mode of thinking that makes sense of the world by integrating individual occurrences into larger schemes of developmental processes, nothing, in principle, fell out of its range. Everything, and literally everything, irrespective of whether it concerned the human or the natural world, could be subjected to the operation of historicization: everything could be made sense of as being prone to change and as being the product of developmental processes that bring together change over time on the one hand (in terms of stages of development) and a deep temporal continuity (in the substance of development) underlying all changes on the other.

In some sort of a distribution of work, in about the same broadly construed time period around the end of the eighteenth and the beginning of the nineteenth century,[8] the historicization of planet Earth has become the task of geology, the historicization of the life on planet Earth has become associated with biology, while the historicization of human life and the social world has become the domain of the professionalized discipline of history (and, with some delay, of sociology). While geology told the story of Earth history through periods just like the discipline of history told the story of human societies, biology conceived of life in terms of evolutionary processes. Both made sense of their object of study along the same developmental temporality as history did with respect to the historicized human world.

The recognition that historical thinking and the operation of historicization is not limited to the human world is gaining momentum lately. Against the backdrop of the long reign of disciplinary codes and epistemological foundations, historians and theorists of history have begun to revisit the scope of historicization. Perhaps it is not too far-fetched to suggest that this growing momentum has something to do with the increasing recognition of the entanglement of the human, technological, and natural worlds. Such recognition may be the impulse behind studies that intend to illuminate the deep connection that binds together modern knowledge production in a shared temporality by which both the human and the natural sciences attempted to make sense of the world. Christophe Bouton claims along these lines that the concept of "development" might have been transferred to historical time from already existing conceptualizations of biological time,[9] while David Schulz argues that geological time had a profound impact on the development of historical thinking and the professionalization of the discipline of history.[10] What is more, those who previously thought that the operation of historicization is confined to the human world tend to reconsider their position lately. On this note, let me quote Frederick Beiser a bit more extensively:

> In stressing the need for a broad definition of historicization – one that includes the natural and human sciences – I depart from Ernst Troeltsch's definition, which I too readily followed in my *The German Historicist Tradition*. According to Troeltsch, the attempt to historicize our thinking is limited to "all our thinking about man, his culture, and sciences." The problem with restricting the definition of historicization to the human sphere is that we ignore the parallel program in the natural sciences. We then blind ourselves to two important facts. First, some of the most stunning results of historicization are found in the natural sciences. I know of no better example of historicization than Darwin's thesis in Chapter XIII of the *Origin of Species* that all true biological classification is "genealogical," i.e., based on lines of descent rather than "some unknown plan of creation." Second, historicization in the natural sciences has sometimes had a profound impact on the human sciences. Consider, for instance, the influence of Kant's *Allgemeine Weltgeschichte und Theorie des Himmels* on the young Herder, who applied the lessons he learned from Kant's cosmology to literature and language.[11]

It cannot be emphasized enough that reconsiderations like this happen in an intellectual atmosphere already occupied with a transdisciplinary debate on the Anthropocene. Although it is not my task here to meticulously argue for a direct causal link between the Anthropocene debate and efforts to recalibrate the scope of historical thinking, it seems rather evident that such present-day issues provide at least an impetus for addressing the question of the limits of our historical sensibility (or their lack thereof). Besides, the Anthropocene debate perfectly testifies to a recent resurgence of history as a mode of thinking across the disciplinary landscape. It is time to introduce the notion properly.

Historical Thinking Across Disciplines

The notion of the Anthropocene emerged in the context of Earth system science. It intends to capture the extent to which anthropogenic changes take place in the integrated system of the Earth and, simultaneously, to denote a new geological epoch in Earth history as indicated by stratigraphic evidence of such human-induced changes in the Earth viewed as a single system of physical and social processes. Only two decades prior to Grimes giving the title *Miss Anthropocene* to her fifth studio album, the notion has been put into wider circulation by Paul Crutzen's two brief articles, the first co-authored with Eugene Stoermer in 2000 and the second published in *Nature* in 2002.[12] Enhancing the inextricable merging of the human and the natural worlds into a concept kicked off two decades of intense discussion of the Anthropocene, practically across the entire scholarly landscape. With respect to the question of disciplinary work, two tendencies can be distinguished in the debate. On the one hand, the notion facilitated the recognition both in the humanities and the social sciences and in the natural and life sciences that there is a necessity to jointly develop conceptual tools adequate for the task of approaching the intertwinement of changes in the human and the natural worlds.[13] On the other hand, the recognition of joint work is just as often accompanied (in weaker versions) and opposed (in stronger versions) by a tendency in the humanities to offer alternatives to what they conceive of as a reigning scientific understanding of the Anthropocene.[14]

The position of this book affiliates with the former tendency aiming at joint work. It seems to me that the conditions of possibility for such collaboration are already provided by the fact that both nature and the realm of human life have been profoundly historicized in the last two centuries or so. Historians who recognize the necessity of joint work and try to overwrite disciplinary codes have good arguments to support their view. Libby Robin at least rather self-evidently remarks right at the onset of her article on the role of history in the Anthropocene that "historians are not alone in studying the past": "as time-scales get longer and documents give way to different kinds of archives, scientific enquiry takes over the study of the past, and *histories* are written by scientists."[15] According to Robin, by compelling us to consider "the role of humanity as a historical force for change in planetary systems," the Anthropocene "demands both geological and historical time-scales, and writes planetary and human histories together."[16] And that could hardly be possible by means other than developing trust in the work of one another and aiming at joining humanities and scientific knowledge.

Bringing "planetary and human histories together" in a meaningful way has already become the imperative of Anthropocene-oriented historical scholarship in the last decade or so. As of now, the task nevertheless remains a challenge. Despite the shared historicity pervading geological and historical time, meeting the demand identified by Robin is not as easy as it may sound. Dipesh Chakrabarty's outstanding efforts devoted to situating the internally differentiated and divided human world (as fraught with inequalities) and the universalizing approach of natural histories (that knows the human on the level of a species) speak the best about the conceptual intricacies involved.[17]

Even though the Anthropocene debate entails the perhaps most overwhelming appeal to history as a mode of thinking across the disciplinary landscape, it is only part of a larger current. Equally interesting and spectacular is the way in which humanities scholarship has begun to expand the scope of historical thinking over the nonhuman world. Attempts to write multispecies histories on the basis of a newly theorized species kinship blend various concerns of emerging knowledge formations such as animal studies, environmental humanities, and critical posthumanism.[18] As multispecies approaches typically aim at overcoming what they identify as an anthropocentric worldview, their relation to the inherent universalism of the Anthropocene predicament can be rather ambiguous. More on this soon. Here the important thing to point out is what brings these approaches to a common platform with Anthropocene scholarship, namely, the extension of their scope over the entanglement of the human and natural worlds.

Now, where does all this leave the discipline of history? According to the recent assessment of Marek Tamm, the Anthropocene "called for a revision and surpassing of various deeply rooted distinctions in historical epistemology," out of which Tamm mentions the ones "between natural history and human history, written history and deep history, human history and multispecies history, national history and planetary history."[19] If Tamm is right, then surpassing these distinctions is likely to end disciplinary "historical epistemology" as we know it. But if I am also right in claiming that a historical mode of thinking and a general historical sensibility in principle allows the historicization of nature and the human world alike, then it is equally plausible to claim that the discipline of history has only begun to embrace a fuller potential already granted by historical thinking in Western modernity. "Historical epistemology" is changing either way, but, in my interpretation, leaving behind disciplinary epistemology poses less difficulties as is commonly thought.

Yet things look more difficult concerning methodology and skills. As Robin notes, when it comes to "the Earth's history over four and a half billion years, geological methods better describe time periods and change than the standard tools of the historian."[20] Although historians have developed various methods to approach the human world in the last two centuries, interpreting stratigraphic evidence is a task of a completely different nature than interpreting written archival evidence, pictorial evidence, or conducting oral history research. Not to mention that skills concerning coping with a certain type of evidence have no merit in themselves. They qualify as skills to cope with evidence only as accompanied by a familiarity with the knowledge system within which the evidence in question functions as evidence in the first place. The disciplinary protocols that enable one to cope only with specific types of evidence, I think, provide yet another reason why no disciplinary expertise can handle alone the task of studying the entangled human-technological-natural world.

But to mention more examples, consider the methodological challenge of studying the nonhuman in animal history. The possibility of the endeavor is, again, granted by the fact that the potential extension of the scope of historicization over the nonhuman is decoded in already existing patterns of historical thought. This is the ground on which—notwithstanding the aforementioned claims concerning a non-anthropocentric scholarship—most approaches to the nonhuman animal extend an already keen sense of injustice within the human world over the nonhuman. In other words, they approach the nonhuman world with a very human agenda of liberation and emancipation. Writing the animal into history is an undertaking that follows the pattern of previous waves of liberation scholarship that aimed at writing the underprivileged and marginalized into history. Developing a truly non-anthropocentric approach would arguably require far more than this, namely, a method that enables the animals to speak for themselves.

So what if enabling animals to speak in a non-anthropocentric approach was theorized as a methodologically feasible enterprise? Even then, serious questions would remain unanswered. Would it be possible, for instance, to disregard the philosophical difficulty that even a method designed to enable animals to speak would still necessarily be human knowledge production about the nonhuman? Or is there a way in which it does not appear as anthropocentric that a *human* method *enables* the animal to speak? In light of its own principles, emancipatory scholarship is supposed to find such an undertaking to be a rather paternalistic one. Perhaps it would be possible then to write history from the animal point of view at least as an unfulfillable desire, as more adventurous approaches advocate lately.[21] According to Ewa Domanska, however, this may not even be the most challenging question:

> The ultimate challenge of animal history, which makes it promising and future-oriented, is not writing animals into history and treating them as agential beings (which has been the main task in the early stage of the field's development), nor even presenting history from "the animal's point of view," but to consider (and possibly to contribute to advancing) interspecies forms of communication that would allow nonhuman beings to report past events [...]. This would require giving up the human epistemic authority of "writing" about the past and opening the field up to various ways of nonverbal communication that would not privilege human language (and language as a privileged way of communication).[22]

The possibility of nonhuman knowledge and its human accessibility points to an entirely different set of questions than the one discussed above. For inasmuch as knowledge itself is nonhuman, it actually does not even pose an epistemological question *for us*. Nor does the prospect of animals reporting past events pose a methodological question, precisely because it does not concern human knowledge production. Instead, as the phrase "epistemic authority" indicates, the question raised by Domanska's scenario is the question of power. And if the "epistemic authority" of "writing" could be given up by humans in the first place, then this would lead to an interspecies translation problem. Yet, at a more profound level, even the assumption that animals would be happy to "report past events" is nothing other than the manifestation of our "epistemic authority." Do animals want to "communicate" history; do they really want to be engaged in our very human endeavor of historical thinking? I have no idea. But it seems that we want them to do so, and I do not see how this does not fall back to anthropocentrism and how such anthropocentrism on the most profound level in human endeavors could be overcome.

Chapter 4 will return to anthropocentrism and related matters in more detail. Although these questions are extremely important for approaches to the more-than-human world, they lead very far from the purposes of this chapter. They are raised here only in order to provide an insight into the perplexities of an already tangible tendency of humanities scholarship in general and historical studies in particular: the growing engagement in constituting the field of human-nature entanglement as an object of study through the extension of the scope of historical thinking and the renegotiation of the limits of historical epistemology.

A New Kind of Historical Thinking

Intricacies and difficulties aside, this appears to be a success story for history and historical thinking so far. It seems that historical thinking has never been more indispensable and relevant than in our times. In yet another perplexing twist, however, this could not possibly be farther from what is actually happening.

As I am writing the first draft of this chapter in the early months of 2019, every day in my Twitter feed looks like a doomsday for the discipline in particular and historical thinking in general. Lamenting over recent phenomena in magazine articles and columns such as the decline in history enrollments appears as if it was a fashionable genre in its own right for pundits and university professors alike.[23] Popular complaints, I think, tend to make hasty connections between particular and general phenomena. On the particular side, there is an educational problem and the question of university funding policy in the US context. On the general side, there is the global loss of the public relevance of the discipline, the supposed lack of historians' public engagement, and the decline of thinking historically in the wider society on the other. In their responses over social media, historians do not seem to fare much better. Although they rightly point at existing forms of public engagement, they seem more interested in defending their profession than to explore a more profound level of the crisis or to question the premise of popular criticism that infers as assumed crisis of "historical thinking" from a recent crisis in history education.[24]

Another line of the defense of history and historical thinking takes the debate out of the context of enrollment statistics and forms of public engagement and focuses on a contemporary social, cultural, and political scene. The recent book by David Andress entitled *Cultural Dementia* is illustrative in this respect. It represents a larger tendency that counterbalances popular complaints about historians' lack of societal engagement by striking back and blaming societal sentiments instead ("cultural dementia"), sort of counter-complaining about the lack of investment in historical thinking in the age of Brexit and Trump.[25] Pointing fingers, however, may not be the most constructive attitude. A more sophisticated self-assurance can be found in Lynn Hunt's argument that the recent flood of lies and fake news call for a reinvestment in historical truth.[26] Either way, the discussion on the relevance of history typically revolves around small-scale recent socio-political occurrences—Trump, Brexit, fake news—without addressing large-scale issues on a deeper level.

This is not to say that paying attention to recent socio-political events in trying to restore the reputation of history is unimportant. The point I want to make concerns only what appears to me as the shortsightedness of a conversation that is supposed to discuss the shape of historical thinking in general. Neither social media debates on the most recent crisis of history nor historians' attempts at self-justification address the big picture, the new constellation of the entanglement of the world of human affairs and nature through unparalleled achievements in the technological domain. This is why, from the viewpoint of this book, everyone involved in the debate appears to be right and wrong at the same time. On the one hand, critics may be right that historical thinking is exhausted and appears less relevant than before—but this is old news. On the other hand, the defense of historians may equally be right in that there is a newly found necessity for such thinking—but the context of this necessity is much broader than contemporary politics and education, and it does not even mean the invocation of the very same historical thinking that is otherwise exhausted.

Let me explain briefly what I mean through quoting—somewhat extensively again—the opening words of Ernst Bresiach's seminal study in historiography. As mentioned earlier, Breisach understands history as the study of change and continuity in the human world. The very first paragraphs of his book, however, are devoted to the growing irrelevance and the perceived exhaustion of historical thinking in his own time (the book was first published in 1983):

> During the nineteenth century historians counseled kings, were leaders in the unification of Germany and Italy, gave a prime minister and a president to France, provided identities to new and old nations, inspired the young American nation in its mastery of a continent, endowed revolutions with the authority of the past, and ascended to the rank of scientists. Above all, they convinced most scholars that everything must be understood in terms of development; in short, historically. No wonder that Thomas Carlyle proclaimed history to be immortal: "Some nations have prophecy, some

have not: but of all mankind, there is no tribe so rude that it has not attempted History, though several have not arithmetic enough to count Five."

Today many smile not just about Carlyle's quaint language, but also about his cocksure confidence. Living in this skeptical age they miss in the passage a proper measure of doubt and caution, if not a share of their suspicion that history has become a bit old-fashioned. Had not the historians of the nineteenth century proclaimed that everybody and everything changes and that there are no timeless concepts? Could it be then that history's days have faded with those of the nineteenth century? Our age, these skeptics argue, may simply require new methods for and new approaches to the "final" explanation of human life or, as some would put it, new intellectual instruments for mastering the world; a world in which it no longer suffices to observe "how things had gradually come to be," as traditional historians have been doing, but one in which historians have to be content with unearthing the raw materials for the social scientists who alone explain, maybe even reorder, human life in a "scientific manner."[27]

Far too many things could be commented on here. I would nevertheless limit myself to mentioning two interrelated aspects. First, it is striking to see that even though these words are less than 40 years old, much of them would come out today as either nonsense or offensive. As to the nonsense part, the skeptics about history demanding a "final" explanation of human life have not much to do with today's skeptics. As to the offensive, Carlyle's words quoted by Breisach—and especially Breisach's lack of distancing—would most likely result in social media storms pointing out how all this sounds not just "quaint" to be smiled upon but Eurocentric and racist. Whether or not social media storms exhibit an attitude to be welcomed is of course an open question. But the fact remains that within a few decades that elapsed since the first edition of Breisach's book, we find ourselves equipped with standards simply incompatible with the standards of the age in which Breisach's book was printed (not to mention Carlyle's age).

Second, this nevertheless proves Breisach's general remarks concerning a growing skepticism about the roots of the kind of a historical thinking that explains the world in terms of developments and "how things had gradually come to be." The fact that we may find no connection today even to the standards of the age of Breisach's book means that we cannot even explain how our current standards might have evolved out of those standards along which Breisach's anecdote could be printed in the first place. Today's scholarly communities claim no association with those standards, and association with the past is the prerequisite of conceiving the present and the future in terms of developments that improve upon pre-existing potentials.

The lack of association with the past and the unfeasibility of understanding the world in terms of developmental processes is precisely the extent to which historical thinking looks exhausted today, on a deeper level than that of contemporary politics and education. This is the extent to which today's complaints about history and historical thinking are right, even if they are right without knowing the reason why. Yet the exhaustion of historical thinking is the exhaustion only of a certain way to make sense of change over time: in terms of developmental processes as confined to the world of human affairs. It is the exhaustion of the very idea of history Breisach insisted on following in his thinking. But historical thinking is itself prone to change. It should not be taken for granted that modern historical thinking remains the same in its essential features since the late eighteenth century, or, let alone, that it stays with us forever in one form or another. Nor is it set to stone that the discipline of history must be the scholarly guardian of historical thinking.

If there is an appeal to history inherent in the growing awareness that human activity brings forth transformations on a planetary scale and in the context of an entangled human-technological-natural world, it is an appeal to a novel kind of historical thinking we need to explicate in the course of our efforts to meet the challenge. As I have argued elsewhere,[28] since the middle of the last century, Western societies increasingly conceive of change over time in other than processual and developmental terms in certain domains of life and human endeavors. I dubbed the new kind of change "unprecedented change" and referred to a new kind of historical thinking that conceives of the world in terms of unprecedented changes as an emerging "evental historical sensibility." The rather spectacular public role of historians in the nineteenth century that Breisach mentions was typical of an age deeply invested in developmental historical thinking concerning socio-political changes in the human world. In contrast, unprecedented change and evental historical sensibility are notions that intend to capture a way of thinking historically that concerns transformations in the entangled human-technological-natural world, with no disciplinary expertise that could guarantee a special public relevance.

All in all, the message this chapter hopes to convey is that the character of the public relevance of historical thinking of our times cannot be boiled down to disciplinary engagements. The appeal to history as a mode of thinking in recent challenges to make sense of anthropogenic planetary change entails a simultaneous appeal—and even a demand—to develop a new kind of historical thinking tailored to investigating a more-than-human world.[29] A proficiency in such a new historical thinking must be broadly construed and must come in a potentially novel shape that cuts across disciplinary boundaries. This book intends to contribute to building such a proficiency by arguing for the necessity to create conceptual tools of a transdisciplinary nature. Concepts that resonate across disciplines may be the ones that enable us to come to terms with the overall constellation of the human-technology-nature entanglement that the next chapter is about to sketch.

Notes

1. Karl Popper's criticism of predicting the course of affairs is perhaps the most widely known classic of the genre. See Karl Popper, *The Poverty of Historicism* (London and New York: Routledge, 2002). For a more recent argument claiming that—even though scientific prediction of the human affairs may be unfeasible—scenario building may enable writing methodologically feasible histories of the future, see David J. Staley, *History and Future: Using Historical Thinking to Imagine the Future* (Plymouth: Lexington, 2010).
2. Marc Bloch, *The Historian's Craft*, trans. Peter Putnam (Manchester: Manchester University Press, 1992), 23.
3. Ernst Breisach, *Historiography: Ancient, Medieval and Modern* (Chicago: University of Chicago Press, 1994), 2.
4. Sarah Maza, *Thinking About History* (Chicago: University of Chicago Press, 2017), 4.
5. R. G. Collingwood, *The Idea of History*, edited with an introduction by Jan van der Dussen (Oxford: Oxford University Press, 1994).
6. Giambattista Vico, *The New Science of Giambattista Vico*, translated from the third edition (1744), trans. Thomas Goddard Bergin and Max Harold Fisch (Ithaca: Cornell University Press, 1948).
7. Dipesh Chakrabarty, "The Climate of History: Four Theses," *Critical Inquiry* 35, no. 2 (2009): 197–222, see esp. 201–207.
8. This is the period that Reinhart Koselleck referred to as *Sattelzeit* (a saddle period), during which the modern temporalized notion of history came to exist. See Reinhart Koselleck, *Futures Past: On the Semantics of Historical Time*, trans. Keith Tribe (New York: Columbia University Press, 2014). More on the *Sattelzeit* in the coda.
9. Christophe Bouton, "From Biological Time to Historical Time: The Category of 'Development' (*Entwicklung*) in the Historical Thought of

Herder, Kant, Hegel, and Marx," in *Biological Time, Historical Time: Transfers and Transformations in 19th Century Literature*, eds. Niklas Bender and Gisèle Séginger (Leiden: Brill, 2018), 61–76.

10. David Schulz, "'O Man! Wilt Thou Never Conceive That Thou Art but an Ephemeron?': The Reception of Geological Deep Time in the Late 18th Century," in *Biological Time, Historical Time: Transfers and Transformations in 19th Century Literature*, eds. Niklas Bender and Gisèle Séginger (Leiden: Brill, 2018), 77–92.
11. Frederick C. Beiser, "Historicization and Historicism: Some Nineteenth Century Perspectives," in *Historisierung: Begriff – Geschichte – Praxisfelder*, eds. Moritz Baumstark und Robert Forkel (Stuttgart: J. B. Metzler Verlag, 2016), 44.
12. Paul J. Crutzen and Eugene F. Stoermer, "The 'Anthropocene,'" (2000), *Global Change Newsletter* no. 41 (May 2000): 17–18; Paul J. Crutzen, "Geology of Mankind," *Nature* 415 (2002): 23.
13. See, for instance, Robert Costanza et al., "Developing an Integrated History and Future of People on Earth (IHOPE)," *Current Opinion in Environmental Sustainability* 4, no. 1 (2012): 106–114; Paul Holm et al., "Collaboration between the Natural, Social and Human Sciences in Global Change Research," *Environmental Science & Policy* 28 (2013): 25–35; Eva Lövbrand et al., "Who Speaks for the Future of Earth? How Critical Social Science Can Extend the Conversation on the Anthropocene," *Global Environmental Change* 32 (2015): 211–218; Libby Robin, "Environmental Humanities and Climate Change: Understanding Humans Geologically and Other Life Forms Ethically," *WIREs Climate Change* 9 (2018): e499.
14. I will return in more detail to the question of the humanities reception of the Anthropocene and the tendency to offer alternatives to the scientific understanding in the next chapter. The alternatives include the Capitalocene, the Technocene, the Chthulucene, the Plantationocene, or the Anthrobscene. See, respectively, Jason W. Moore, "The Capitalocene, Part I: On the Nature and Origins of Our Ecological Crisis," *The Journal of Peasant Studies* 44, no. 3 (2017): 594–630; Alf Hornborg, "The Political Ecology of the Technocene: Uncovering Ecologically Unequal Exchange in the World-System," in *The Anthropocene and the Global Environmental Crisis: Rethinking Modernity in a New Epoch*, eds. Clive Hamilton, François Gemenne and Christophe Bonneuil (London and New York: Routledge, 2015), 57–69; Donna J. Haraway, *Staying with the Trouble: Making Kin in the Chthulucene* (Durham: Duke University Press, 2016); Donna Haraway et al., "Anthropologists Are Talking – About the Anthropocene," *Ethnos* 81, no. 3 (2016): 535–564; Jussi Parikka, *The Anthrobscene* (Minneapolis: University of Minnesota Press, 2014).

15. Libby Robin, "Histories for Changing Times: Entering the Anthropocene?" *Australian Historical Studies* 44, no. 3 (2013): 329.
16. Robin, "Histories for Changing Times," 329.
17. For Chakrabarty's first essay on the Anthropocene and its challenge to historical thinking, see Chakrabarty, "The Climate of History." For his latest major intervention, see Dipesh Chakrabarty, "Anthropocene Time," *History and Theory* 57, no. 1 (2018): 5–32. For more pioneer scholarship with respect to the question of history and the Anthropocene, see Libby Robin and Will Steffen "History for the Anthropocene," *History Compass* 5, no. 5 (2007): 1694–1719; Julia Adeney Thomas, "History and Biology in the Anthropocene: Problems of Scale, Problems of Value," *American Historical Review* 119, no. 5 (2014): 1587–1607.
18. See, for instance, Donna J. Haraway, *When Species Meet* (Minneapolis: University of Minnesota Press, 2008); Anna Lowenhaupt Tsing, *The Mushroom at the End of the World: On the Possibility of Life in Capitalist Ruins* (Princeton: Princeton University Press, 2015); Stacy Alaimo, *Exposed: Environmental Politics and Pleasures in Posthuman Times* (Minneapolis: University of Minnesota Press, 2016).
19. Marek Tamm, "Introduction: A Framework for Debating New Approaches to History," in *Debating New Approaches to History*, eds. Marek Tamm and Peter Burke (London: Bloomsbury, 2018), 7.
20. Robin, "Histories for Changing Times," 329.
21. Cf. Erica Fudge, "What Was It Like to Be a Cow? History and Animal Studies," in *The Oxford Handbook of Animal Studies*, ed. Linda Kalof (New York: Oxford University Press, 2017), https://doi.org/10.1093/oxfordhb/9780199927142.013.28; see also David Gary Shaw, "A Way with Animals," *History and Theory* 52, no. 4 (2013): 1–12.
22. Ewa Domanska, "Animal History," *History and Theory* 56, no. 2 (2017): 285.
23. The two articles that have been popping up most frequently in my Twitter feed are Eric Alterman, "The Decline of Historical Thinking," *The New Yorker*, 4 February 2019. https://www.newyorker.com/news/news-desk/the-decline-of-historical-thinking; and Max Boot, "America's Ignorance of History is a National Scandal," *The Washington Post*, 20 February 2019. https://www.washingtonpost.com/opinions/americans-ignorance-of-history-is-a-national-scandal/2019/02/20/b8be683c-352d-11e9-854a-7a14d7fec96a_story.html?utm_term=.0749d635325b
24. On a side-note, whereas debating the drop in history enrollments generalizes into a crisis of "historical thinking," the debate typically does not even address the most approximate general context of history education: the

potential downfall of universities as bastions of knowledge production. The latter, I think, is a large-scale phenomenon that should actually be worthy of debate.

25. David Andress, *Cultural Dementia: How the West Has Lost Its History, and Risks Losing Everything Else* (London: Head of Zeus, 2018).
26. Lynn Hunt, *History: Why It Matters* (Cambridge: Polity, 2018).
27. Breisach, *Historiography*, 1.
28. Zoltán Boldizsár Simon, *History in Times of Unprecedented Change: A Theory for the 21st Century* (London: Bloomsbury, 2019).
29. For more details, see Marek Tamm and Zoltán Boldizsár Simon, "More-than-Human History: Philosophy of History at the Time of the Anthropocene," in *Philosophy of History: Twenty-first-century Perspectives*, ed. Jouni-Matti Kuukkanen (London: Bloomsbury, 2020), forthcoming.

Bibliography

Alaimo, Stacy. *Exposed: Environmental Politics and Pleasures in Posthuman Times.* Minneapolis: University of Minnesota Press, 2016.

Alterman, Eric. "The Decline of Historical Thinking," *The New Yorker*, 4. February 2019. https://www.newyorker.com/news/news-desk/the-decline-of-historical-thinking

Andress, David. *Cultural Dementia: How the West Has Lost Its History, and Risks Losing Everything Else.* London: Head of Zeus, 2018.

Beiser, Frederick C. "Historicization and Historicism: Some Nineteenth Century Perspectives," in *Historisierung: Begriff – Geschichte – Praxisfelder*, edited by Moritz Baumstark und Robert Forkel, 42–54. Stuttgart: J. B. Metzler Verlag, 2016.

Bloch, Marc. *The Historian's Craft.* Translated by Peter Putnam. Manchester: Manchester University Press, 1992.

Boot, Max. "America's Ignorance of History is a National Scandal," *The Washington Post*, 20. February, 2019. https://www.washingtonpost.com/opinions/americans-ignorance-of-history-is-a-national-scandal/2019/02/20/b8be683c-352d-11e9-854a-7a14d7fec96a_story.html?utm_term=.0749d635325b

Bouton, Christophe. "From Biological Time to Historical Time: The Category of "Development" (*Entwicklung*) in the Historical Thought of Herder, Kant, Hegel, and Marx." In *Biological Time, Historical Time: Transfers and Transformations in 19th Century Literature*, edited by. Niklas Bender and Gisèle Séginger, 61–76. Leiden: Brill, 2018.

Breisach, Ernst. *Historiography: Ancient, Medieval and Modern.* Chicago: University of Chicago Press, 1994.

Chakrabarty, Dipesh. "The Climate of History: Four Theses." *Critical Inquiry* 35, no. 2 (2009): 197–222.

Chakrabarty, Dipesh. "Anthropocene Time." *History and Theory* 57, no. 1 (2018): 5–32.

Collingwood, R. G. *The Idea of History*, edited with an introduction by Jan van der Dussen, Oxford: Oxford University Press, 1994.

Costanza, Robert et al. "Developing an Integrated History and Future of People on Earth (IHOPE)." *Current Opinion in Environmental Sustainability* 4, no. 1 (2012): 106–114.

Crutzen, Paul J. "Geology of Mankind." *Nature* 415 (2002): 23.

Crutzen, Paul J., and Eugene F. Stoermer. "The 'Anthropocene.'" *Global Change Newsletter* no. 41 (2000): 17–18.

Domanska, Ewa. "Animal History." *History and Theory* 56, no. 2 (2017): 267–287.

Fudge, Erica. "What Was It Like to Be a Cow? History and Animal Studies." In *The Oxford Handbook of Animal Studies*, edited by Linda Kalof, New York: Oxford University Press, 2017. Available at: https://doi.org/10.1093/oxfordhb/9780199927142.013.28

Haraway, Donna J. *When Species Meet.* Minneapolis: University of Minnesota Press, 2008.

Haraway, Donna J. *Staying with the Trouble: Making Kin in the Chthulucene.* Durham: Duke University Press, 2016.

Haraway, Donna et al. "Anthropologists Are Talking – About the Anthropocene." *Ethnos* 81, no. 3 (2016): 535–564.

Holm, Paul et al. "Collaboration between the Natural, Social and Human Sciences in Global Change Research." *Environmental Science & Policy* 28 (2013): 25–35.

Hornborg, Alf. "The Political Ecology of the Technocene: Uncovering Ecologically Unequal Exchange in the World-System." In *The Anthropocene and the Global Environmental Crisis: Rethinking Modernity in a New Epoch*, edited by Clive Hamilton, François Gemenne and Christophe Bonneuil, 57–69. London and New York: Routledge, 2015.

Hunt, Lynn. *History: Why It Matters.* Cambridge: Polity, 2018.

Koselleck, Reinhart. *Futures Past: On the Semantics of Historical Time.* Translated by Keith Tribe. New York: Columbia University Press, 2014.

Lövbrand, Eva et al. "Who Speaks for the Future of Earth? How Critical Social Science Can Extend the Conversation on the Anthropocene." *Global Environmental Change* 32 (2015): 211–218.

Maza, Sarah. *Thinking About History.* Chicago: University of Chicago Press, 2017.

Moore, Jason W. "The Capitalocene, Part I: On the Nature and Origins of Our Ecological Crisis." *The Journal of Peasant Studies* 44, no. 3 (2017), 594–630.

Parikka, Jussi. *The Anthrobscene.* Minneapolis: University of Minnesota Press, 2014.

Popper, Karl. *The Poverty of Historicism.* London and New York: Routledge, 2002.

Robin, Libby. "Histories for Changing Times: Entering the Anthropocene?" *Australian Historical Studies* 44, no. 3 (2013): 329–340.

Robin, Libby. "Environmental Humanities and Climate Change: Understanding Humans Geologically and Other Life Forms Ethically." *WIREs Climate Change* 9 (2018): e499.

Robin, Libby, and Will Steffen "History for the Anthropocene." *History Compass* 5, no. 5 (2007): 1694–1719.

Schulz, David. "'O Man! Wilt Thou Never Conceive That Thou Art but an Ephemeron?': The Reception of Geological Deep Time in the Late 18th Century." In *Biological Time, Historical Time: Transfers and Transformations in 19th Century Literature*, edited by Niklas Bender and Gisèle Séginger, 77–92. Leiden: Brill, 2018.

Shaw, David Gary. "A Way with Animals." *History and Theory* 52, no. 4 (2013): 1–12.

Simon, Zoltán Boldizsár. *History in Times of Unprecedented Change: A Theory for the 21st Century*. London: Bloomsbury, 2019.

Staley, David J. *History and Future: Using Historical Thinking to Imagine the Future*, Plymouth: Lexington, 2010.

Tamm, Marek. "Introduction: A Framework for Debating New Approaches to History." In *Debating New Approaches to History*, edited by Marek Tamm and Peter Burke, 1–19. London: Bloomsbury, 2018.

Tamm, Marek, and Zoltán Boldizsár Simon. "More-than-Human History: Philosophy of History at the Time of the Anthropocene." In *Philosophy of History: Twenty-first-century Perspectives*, edited by Jouni-Matti Kuukkanen, forthcoming, London: Bloomsbury, 2020.

Thomas, Julia Adeney. "History and Biology in the Anthropocene: Problems of Scale, Problems of Value." *American Historical Review* 119, no. 5 (2014): 1587–1607.

Tsing, Anna Lowenhaupt. *The Mushroom at the End of the World: On the Possibility of Life in Capitalist Ruins*. Princeton: Princeton University Press, 2015.

Vico, Giambattista. *The New Science of Giambattista Vico*. Translated from the third edition (1744). Translated by Thomas Goddard Bergin and Max Harold Fisch. Ithaca: Cornell University Press, 1948.

CHAPTER 3

The Entangled Human-Technological-Natural World

Abstract This chapter argues that the entanglement of the human, the technological, and the natural worlds constitutes a new object of knowledge. The focus of the chapter lies with the Anthropocene debate, paying special attention to two themes: first, the role of technology in the overall picture, and second, the question concerning the form of joint work across disciplines that may be necessary for coping with the new predicament. As to the first, the chapter argues that technology is a systemic element of the entanglement; as to the second, it criticizes a tendency in humanities scholarship to initiate parallel discussions to scientific ones (leading to a *runaway agonism* in Anthropocene research) and calls for developing *connective concepts* instead. The relevance of the chapter for the argument of the book is that a new kind of epochal thinking emerges against the backdrop of the human-technology-nature entanglement.

Keywords Technology • Human • Nature • Entanglement • Anthropocene • Connective concepts

Technology as a Systemic Element

The Anthropocene is the most successful notion to capture an overall constellation of the entanglement of the human, the technological, and the natural worlds. As mentioned in the previous chapter, it emerged

Z. B. Simon, *The Epochal Event*, Palgrave Studies in the History of Science and Technology,
https://doi.org/10.1007/978-3-030-47805-6_3

within Earth system science, referring to anthropogenic changes in the Earth system and naming a potential geological epoch in which human activity transforms the Earth system as is attested to by stratigraphic evidence. Following the popularization of the notion in the first years of the new millennium, the wider discussion within Earth system science and the stratigraphic community led to the establishment of the Anthropocene Working Group in 2009, assigned with the task of investigating the question whether the Anthropocene should be formalized as part of the geologic time scale.

The preliminary findings and recommendations of the Anthropocene Working Group were presented in 2016 at the 35th International Geological Congress in Cape Town, South Africa, and published as an article the following year. The recommendations represent the results of an internal voting process on several interrelated questions, involving the 35 working group members. To briefly recapitulate the results: the Anthropocene is "stratigraphically real" (34 votes in favor); it should be formalized (30 in favor); the formalization should mean the hierarchical level of an epoch (20.5 in favor, other options being era, period, subepoch, and sub-age); with the onset around 1950 (28.5 in favor); in which the onset should be defined by a "golden spike," that is, by a Global boundary Stratotype Section and Point (GSSP) meaning stratigraphic evidence (25.5 in favor) as opposed to being defined by a Global Standard Stratigraphic Age (GSSA) meaning a calendar date; and the best "primary marker" of the epoch is plutonium fallout (10 in favor, out of 12 available options).[1]

Now, how exactly does this signify the entanglement of the human world, nature, and the technological domain? And, most importantly, how is technology involved in the overall picture? As the previous chapter already introduced the intertwinement of the human and natural worlds in both Anthropocene scholarship and multispecies approaches, it is the role of technology that demands special attention here. The reason to focus on technology is that the technological component is often underemphasized in the overall picture. Maybe the role of technology in the human-nature entanglement is simply taken for granted, or perhaps the entire technological domain is considered to be part of the human world by default. Either way, it is of utmost importance to stress how the recent entanglement of the human and the natural worlds is tied with technology. For, as Dolly Jørgensen empathically puts it, anthropogenic transformations in the Earth system "may be at the hands of humans, but they are not because of human hands – they are the result of human technologies."[2]

Jørgensen's words remind us that the predicament we are struggling to grasp is not an equation of two variables, thus it should not be restricted to a human-nature binary framework. Returning to the recommendations of the Anthropocene Working Group with these considerations in mind, it is not hard to see how voting for plutonium fallout as the "primary marker" of the Anthropocene already assigns a central—although implicit—importance to technology.[3] It conveys the message that a particular technological "achievement," that is, the development and use of nuclear weaponry, marks a new epoch on a geological scale. And if this sounds too particular or specific, consider the fact that the case would be the same even if the members voted differently. For most of the options they voted for as primary markers are products of the technological developments of industrial modernity, or of the post-Second World War period. The votes for plastic, fuel ash particles, carbon dioxide concentration, or radiocarbon bomb spike as primary markers are all votes for technology-produced records in Earth's strata.[4] The last one, referring to the peaking of ^{14}C in the atmosphere as the result of atmospheric nuclear weapons testing, is even connected to the very same technological development as the most voted option.

This does not mean that there has been no human impact on the Earth system before industrial modernity or the mid-twentieth-century developments in nuclear technology. It means only two things with respect to the vocabulary of "impact" in relation to the question of modern technology and the notion of the Anthropocene. First, when the Anthropocene Working Group argues that human activity had an effect on the Earth system that is "both large and synchronous" only as beginning in the mid-twentieth century,[5] then what they emphasize are the multiple and interrelated consequences of technology. What they are seeking are not individual "impacts" but a systemic human intervention into Earth system processes through technology. The scale and scope of such human activity is best captured by the Great Acceleration thesis, which also argues for a postwar onset of the Anthropocene by pointing at the simultaneous runaway of several Earth system indicators and socio-economic trends.[6]

Second, attempting to link a loosely defined notion of "human impact" to the Anthropocene predicament—so loosely defined that it allows to assess each individual "human impact" as constitutive of the Anthropocene predicament itself—comes at a price of downplaying the role of modern technology. Consider the argument about how the Great Dying in the Americas as a consequence of European expansion had an "impact" on the Earth system. On the one hand, the argument brings humanities scholarship to the fore in the general debate; on the other hand, linked with a position that sets the beginning of the Anthropocene in 1610,[7] it necessarily disconnects the Anthropocene predicament from the kind of modern technology that explicitly attempts to kick off and steer natural processes (whether successfully and desirably or not is another question).

At best, the notion of technology that can be linked to an understanding of an early Anthropocene implies an understanding of technology just as broad as the matching understanding of Anthropocene. One can follow, for instance, Eric Schatzberg's recent history of the concept of technology that advocates the reinstitution of the understanding of technology "as an expression of human culture, as art in the older sense of the term, as a union of thought and action, mind and hand," against the more recent understanding of technology as applied science.[8] But, again, the risk in such a broad understanding consists of losing sight of the specific technology that entangles the human and the natural words on a systemic level. What if this specific technology is linked to the understanding of technology as applied science? What if old understandings of technology do not enable us to understand the novelty of our situation? Does not the postwar spread of the understanding of technology as applied science correlate with the "large and synchronous" traces of human activity left in the natural strata? And if yes, is not this somehow indicative of the kind of technology that can be linked with the Anthropocene? Why should we understand technology in old ways when, in fact, new conceptions of technology are the ones that shape our lives in unprecedented ways?

When aiming at understanding our situation, the imperative is to get a grip on the workings of the specific notion of technology instrumental in our own times. And the specificity we identify—whether we like or not—should not be explained away. Neither by longing for notions and ideas long abandoned or discredited, nor by making the new to appear to be old. The perils of the former are sketched above, while the chief peril in the latter is that the farther we stretch the Anthropocene back in time the less shocking it occurs to us. The novelty of our present condition fades away with the operation. What leaving a specifically mid-twentieth-century technology out of the equation would mean is best indicated by the following quote from Hannah Arendt:

> Up to our own age human action with its man-made processes was confined to the human world, whereas man's chief preoccupation with regard to nature was to use its material in fabrication, to build with it the human artifice and defend it against the overwhelming force of the elements. The moment we started natural processes of our own – and splitting the atom is precisely such a man-made natural process – we not only increased our power over nature, or became more aggressive in our dealings with the given forces of the earth, but for the first time have taken nature into the human world as such and obliterated the defensive boundaries between natural elements and the human artifice by which all previous civilizations were hedged in.[9]

To see the novelty of our condition, to see its radicality, technology must be part of the entanglement. And not just any technology, but the specific one that marks "the moment we started natural processes of our own." Yet the human-technology-nature entanglement is not restricted to particularities such as splitting the atom mentioned by Arendt or the industrial production of plastic waste that features in the news today. Nor is the technological component the mere aggregate of individual techno-produced records. It is much larger in scale and scope within the broader picture; it is a systemic element.

Alf Hornborg has long advocated such a systemic understanding of technology by discussing a growing technological infrastructure and products in terms of a "technomass," a neologism invented to the analogy of "biomass."[10] Technomass, however, is unevenly produced, which makes Hornborg think that Anthropocene may not be the best term for the predicament to which technology is central. He suggests to refer to it as Technocene instead—a term more attentive to inequalities within the human world.[11] Although it is very unlikely that the Anthropocene would be discarded in favor of the Technocene, the notion of technomass is instrumental in indicating not only that the technological domain is integral to the Anthropocene predicament on a systemic level but also the fact that technology is not an abstract idea but the product of an internally differentiated human world.

The latter point is precisely what seems to be missing from the competing term that nevertheless resonates better with Earth system science and with the work of the Anthropocene Working Group. Peter Haff's notion of the technosphere as a system of the Earth (just as well as the hydrosphere or the biosphere) regards technology to be a systemic element too.[12] Unlike Hornborg's technomass, however, the technosphere is more ambiguous in its relation to the human world. In Haff's understanding, the technosphere includes not only "the world's large-scale energy and resource extraction systems, power generation and transmission systems, communication, transportation, financial and other networks, governments and bureaucracies, cities, factories, farms and myriad other 'built' systems," but also "parts of these systems," such as "computers, windows, tractors, office memos and humans," and "systems which traditionally we think of as social or human-dominated, such as religious institutions or NGOs."[13]

As "a system for which humans are essential but, nonetheless, subordinate parts,"[14] the technosphere is autonomous. Its autonomy means nothing less than abandoning "the apparently natural assumption that the technosphere is primarily a human-created and controlled system and instead develop the idea that the workings of modern humanity are a product of a system that operates beyond our control and that imposes its own requirements on human behavior."[15] Needless to say, these are strong claims, especially to the ears of humanities scholars. But not to Earth system scientists, who bring Haff's conceptual innovation to a study of technofossils (referring to the human-made material remains of the technosphere), aiming at developing technostratigraphic divisions while arguing that "the middle of the 20th century has seen a change from local technostratigraphies to, essentially, a global one, enhancing the potential of this time level as an appropriate and perhaps formal Anthropocene beginning."[16]

Haff's autonomous technosphere and Hornborg's unevenly produced technomass seem to work in different registers. The technological determinism underlying Haff's notion of the technosphere is just as unlikely to resonate with most humanities and social scientific scholarship as improbable it is that Hornborg's Technocene wins over Earth system science and replaces the notion of the Anthropocene. Yet the two views share a common ground in treating technology as a systemic element in an entangled human-technological-natural world. In that, they represent a promising start for developing concepts with a transdisciplinary appeal, which, I believe, is precisely what we need today. Such concepts may help us gaining a better understanding of the overall constellation of the entanglement of the human and natural worlds as arising out of the workings of the increased technological powers that can likely be linked with post-Second World War modes of societal organization—even if, for now, we insufficiently understand the nature of the connection.

It is of course open to debate whether the agenda of achieving at least a very basic shared conceptual understanding across disciplines is desirable in the first place. As mentioned in the previous chapter, the Anthropocene is a contested concept with two major tendencies in its reception in the humanities and the social sciences: one that advocates joint work with the sciences and one that offers alternatives to a scientific understanding of the Anthropocene. The two tendencies are not necessarily contradictory, even if they oftentimes conflict. Besides, the tendencies themselves are not necessarily based on a deep agreement in their respective scopes over the particular forms that either joint work or alternatives to scientific understanding should take.

Within a matrix of many positions, this book advocates the creation of *connective concepts* as a form of joint work that the remaining pages of this chapter try to situate with other views in the Anthropocene debate. To see the significance of connective concepts, let's begin with a brief overview of the available alternatives.

The Entanglement as a New Object of Knowledge

At its most constructive, the humanities and social scientific perspective is invoked to improve upon the way in which "the mainstream story projected by leading environmental scientists" tends to offer "a restricted understanding of the entangled relations between natural, social and cultural worlds."[17] At least this is what Eva Lövbrand and her co-authors emphasize in a paper devoted to demonstrating the potential contributions of social and political science to the Anthropocene debate. The question of the hour is that of how exactly to foster a better understanding of "the entangled relations between natural, social and cultural worlds." It seems reasonable to suppose that the ideal way would entail a two-way transfer of knowledge between the natural and the life sciences on the one side, and the human and social sciences on the other. The imperative of such a mutual knowledge transfer across disciplines might be best captured by

Libby Robin's phrase "understanding humans geologically and other life forms ethically."[18]

With respect to developing an understanding of the entanglement of human and nonhuman worlds, however, approaches to a multispecies world, posthumanist epistemologies, and, generally speaking, the critical humanities scholarship mentioned in the previous chapter—the one that brings socio-political agendas to the overall discussion through extending an attentiveness to inequalities and injustices in the human world to the nonhuman world—may be even better positioned.[19] At the same time, precisely because they conceive of the new challenge as a spin-off of an older demand, they also represent what seems to constitute a less constructive way in which humanities and social scientific scholarship joins the Anthropocene debate. In their reception of the Anthropocene concept, critical humanities scholarship tends to see an undesired return of a universal notion of humanity. It associates the *anthropos* (human in Greek) of the Anthropocene with the modern project of human mastery with the figure of "Man" at its center. Critical humanities scholarship argues that an undifferentiated notion of the human seems to resurface in the scientific Anthropocene discussion, despite the work of feminist, postcolonial, poststructuralist, and deconstructionist critique in the last more than half-century. By bringing the question of values and responsibility to the fore, it interprets the notion of the Anthropocene as one that entails a victimization of the whole of humanity instead of the few who are actually to blame for launching catastrophic planetary futures.

Approaches less concerned with Western universalism and anthropocentrism and more with capitalist modes of production are equally upfront in specifying responsibilities when arguing against the species view of humanity. Consider the following powerful sentence from a joint article by Andreas Malm and Hornborg (an instant classic):

> Capitalists in a small corner of the Western world invested in steam, laying the foundation stone for the fossil economy: at no moment did the species vote for it either with feet or ballots, or march in mechanical unison, or exercise any sort of shared authority over its own destiny and that of the Earth system.[20]

Now, why would this critique appear as less constructive? Isn't it important to make clear that bringing about the current condition is rather obviously not the work of the whole of humanity? Of course it is. But there are at least three respects in which automated humanities and social scientific reflexes may miss the mark. First, scientists seem perfectly aware of inequalities. The ecosocialist Ian Angus (who can hardly be accused of being inattentive to inequalities in the human world) took great pains recently to dissipate the idea of critical humanities that the sciences are blind to the uneven distribution of responsibility.[21] Besides, as Dipesh Chakrabarty and Julia Adeney Thomas have both pointed out recently, the sciences are not invested in "blaming" anyone in the first place.[22]

The second and closely related aspect concerns the focus of the critical stance, and the work of Julia Adeney Thomas is again instructive in briefly sketching it. Thomas and Michael Ellis argue in a co-authored paper that the tendency of humanities and social scientific criticism to reduce attention to climate change and the question of fossil fuels prevents us from seeing the larger picture. They warn against conflating the particular problem of climate change and the systemic Anthropocene condition of anthropogenic changes in the Earth system. As soon as one focuses on the latter—which of course includes climate change and the burning of fossil fuels without being reducible to them—it becomes possible to see how the Anthropocene is not simply a "Western story" and how Asia plays a central role not only in suffering the consequences of the Anthropocene but also in bringing about its systemic condition.[23] To be clear, a more systemic view does not attempt to do away with questions of responsibility. It affirms the necessity of the Earth system science view just in order to be able to adequately situate it with questions of responsibility (instead of simply replacing scientific questions with humanities ones).

Finally, the *anthropos* in the notion of the Anthropocene does not necessarily have much to do with the idea of humanity as deconstructed by humanities and social scientific scholarship in the last half-century, that is, the idea of humanity coming to the fulfillment of its potential over the course of a historical process. The humanity invoked by scientific Anthropocene research is indeed a universal notion, but it may be more of a new universal idea that we are yet to understand and conceptualize and less of the return of an old one. It does not advocate any social, cultural, and political universalism that would gesture toward unifying humanity in said terms, which are the traditional concerns of the humanities and social sciences. Instead, it is structured by the existential risk of a potential self-extinction of a biological species.[24] More on this in the next chapter.

Lacking a constructive attitude may be an obstacle for the human and social sciences to become just as entangled with other approaches as are the human, the technological, and the natural worlds today, potentially constituting a common object of study. In many respects, critical humanities scholarship does not seem to foster a *shared* effort to understand the entangled human-technological-natural world as based on *mutual* knowledge transfer. Oftentimes it points to crucial topics and themes to discuss in meeting new challenges only inasmuch as this enables retaining the discussion of topics, themes, and theoretical frameworks it already had before meeting the very challenges. Consider, for instance, the "critical climate change" project of Tom Cohen and Claire Colebrook, insisting on the textual approach:

> Rather than say that the humanities can save themselves by becoming environmental or that ecocriticism must be the future of a world whose priority ought to be survival, we think of matter as too plural, volatile, and inhuman to provide any such escape. Matter is textual and certainly not something that can justify a posttheoretical literalism, anticritique, or surface reading; the events of climate change and the Anthropocene *intensify* the textual recalcitrance of materiality, precluding any notion of a return to what would supposedly precede and cause the text.[25]

There can hardly be a challenge worthy of its name that only reinforces pre-challenge beliefs. Yet this is what seems to happen in the above quote. The "critical climate change" project does not suggest that a textual approach—trending and becoming institutionalized in the 1980s at Anglo-American universities—may potentially bring some compelling insights to our efforts directed at meeting new challenges raised outside the scope of humanities scholarship. Rather, its priorities lie with making use of a new external challenge in an internal contestation among humanities approaches by claiming that there is either the textualism of humanities critical theory (not to be confused with the critical theory of the Frankfurt School) or "posttheoretical literalism" and "anticritique." The latter terms associate certain recent developments in humanities scholarship with a reactionary pre-theoretical naivety and attempt to make the case thereby for the sustained reign of a once appealing approach designed to meet former (and most certainly pre-Anthropocene) challenges.[26]

To be clear, this is not to say that the approach of Cohen and Colebrook in particular and critical humanities scholarship in general should uncritically approve of the work of the sciences. It is only to say that, for the moment, humanities approaches tend to fail to consider the extent to which their old expertise may also need to change in order to be able to adequately respond to the issues raised by the human-technology-nature entanglement. What is more, humanities views are sometimes expressed in unnecessarily combative ways, even in ways openly hostile to scientific views. Needless to say, this also does not point toward the prospect of joint work.

As an example, one may think of the co-authored book by French historians Christophe Bonneuil and Jean-Baptiste Fressoz, *The Shock of the Anthropocene*. While the book is brilliant in sketching a variety of Anthropocene narratives that bring (modern) historical thinking to the Anthropocene predicament, in its less brilliant moments it depicts scientific work as a means to acquire global governance. Telling what Bonneuil and Fressoz call the "Geocratic Grand Narrative of the Anthropocene" is assumed to be the way in which "anthropocenologists" justify aspirations for a "geo-government of scientists."[27] This almost sounds like a conspiracy theory. Fortunately, there is no need to argue either for how such hostility is misplaced and unjustified or for the equally apparent fact that scientist know their limits and themselves call for humanities and social scientific expertise, because, again, Julia Adeney Thomas already brilliantly argued for both points.[28]

What can nevertheless be indicated here is the Earth system science view on disciplinary inclusivity. Consider the composition of the Anthropocene Working Group, inviting experts from a diverse scholarly background:

> From the beginning, the AWG [Anthropocene Working Group] represented a broader community than is typical of ICS [International Commission of Stratigraphy] working groups, which for the most part consist mostly or entirely of stratigraphers and palaeontologists experienced in the rocks and fossils of the particular time unit under study. However, because the Anthropocene concept not only spans geological time but also involves an evaluation of human impact upon the Earth System through historical and instrumental records, it was considered appropriate to include representatives of the community working on the processes of contemporary global change including climate science, ecology, archaeology, human history and the history of science, oceanography, polar science and even international law.[29]

That said, the focus of the report is admittedly on Earth system science and geological concerns. While scientists doing their job in explicating a concept of their own creation and inviting a broad scholarly community to join the discussion most certainly does not justify hostility, it must equally be pointed out that the inviting gestures can also fall short in facilitating joint work on a substantial level. To begin with, joining the work of the Anthropocene Research Group means voting on the concerns raised and defined by Earth system science. This is not necessarily the kind of inclusivity that would enable much operational space for the humanities and the social sciences. Many strands of humanities and social scientific scholarship would agree that a basis for joint work must arise out of the work of the sciences that conceptualized anthropogenic planetary change in the first place, but proceeding with joint work on that basis should ideally mean more than giving some votes to humanities scholars in discussing matters of Earth system science.

I think that neither the inviting gestures of the sciences nor the parallel discussions of the humanities and the social sciences represent the most fruitful strategies in coping with perceived radical novelty. If we really think that we face a challenge, and if we really think that the challenge is a novel one, then we must acknowledge that the entanglement of the world of human affairs, technology, and the natural world constitute a new object of knowledge. Only insofar as we affirm the constitution of a novel object of knowledge can we approve of a demand of joint work and the necessity to create equally new knowledge formations designed to investigate the new object of study. This does not simply mean that, in order to be able to tackle the questions and problematics arising out of the greatest challenges of our own times, the concerns and modes of expertise of already existing knowledge formations must be brought together in a potentially new knowledge regime. It rather means that affirming and responding to such challenges entail that already existing concerns and modes of expertise may change beyond recognition on both sides. Some may be retained or become integrated into new frameworks, while others may vanish. It is hardly foreseeable how things play out with respect to this. We will have to see while developing our responses in the shape of concepts that somehow stand a chance of finding resonance among the members of both scholarly communities as they exist today.

Runaway Agonism and Connective Concepts

As of now, the frequency of claims that the Anthropocene rewrites existing disciplinary is not matched by the frequency of coming to the recognition that the human-technology-nature entanglement represents a new object of knowledge. Without such a recognition, the possibility of finding a shared basis for joint work is, I think, considerably low.

One of the greatest obstacles to a meaningful transdisciplinary engagement with the Anthropocene predicament today is the fierce contestation over the very meaning of the notion. The aforementioned humanities and social scientific suspicion about the undifferentiated *anthropos* and the resulting parallel discourses represent only one split within Anthropocene scholarship, although quite a spectacular one. The accompanying humanities drive for inventing playful puns on the notion—ranging from the already mentioned Technocene through the Anthrobscene[30] and the Capitalocene to the Plantationocene[31]—will likely not resonate very well with the work of the sciences. Besides, as Ian Angus points out,[32] alternative notions even derive from a misunderstanding concerning the *-cene* suffix, which does not denote an age, era, or epoch. It means "recent" or "new" as deriving from the Greek *kainos*—and this etymological information can be checked even on Wikipedia.

Another instance of a profound fissure lies between bleak and shiny perceptions of planetary-scale human activity. Anthropogenic changes in the Earth system are conceived of as ones that may engender societal collapse at best and ones that result in planetary conditions which no longer support human life on Earth at worst. The drive behind human endeavors that end up transforming the planet is frequently associated today with the larger projects of Western humanism and Western modernity on various levels. Those who approach the Anthropocene in ethical terms, critiquing what they consider as instances of hubristic human entitlement and misguided ideas of human mastery responsible for the dire prospects, have serious difficulties with finding shared concerns with those who put their trust in the very same ideas and human capacities as a means to steer the situation toward a good Anthropocene. Chapter 4 will explore the clash of views on increased human capacities in more detail.

Old conventions may result in further divisions. In a recent article, researchers from Finland distinguish at least four major approaches to the Anthropocene: a geological, a biological, a social, and a cultural one.[33] These disciplinary divisions typically work within the aforementioned larger divide between natural scientific approaches on the one hand and social scientific and humanities approaches on the other. Tero Toivanen and the six co-authors of the article share a common platform with this book in arguing for the necessity of transdisciplinary Anthropocene research. But their understanding of transdisciplinary knowledge production departs from my arguments that the entangled human-technological-natural world as a new object of study demands an equally new knowledge regime attuned to studying the entanglement. They rather think that "the possible antagonisms between disciplines" may represent "fruitful tensions for better future scientific understanding."[34]

When referring to antagonisms, Toivanen et al. likely mean an *agonistic pluralism* instead. In political theory, Chantal Mouffe describes the potentially constitutive conflicts in a we/they relation of adversaries as agonism, while antagonism refers to a we/they relation consisting of two sides that consider each other as enemies.[35] An agonistic pluralism within Anthropocene research—inasmuch as it is opposed to an antagonistic enemy-seeking—is not incompatible with the constellation I would prefer. Not in principle at least. The current state of practical matters is another question, and the prospects in this respect are far less promising than on the level of potential theoretical compatibilities.

Agonism, I think, works until it does not concern the very basic premises of research. The most troubling aspect of today's Anthropocene research is precisely that it does. The severity of the cross-cutting fissures listed above hardly enables us to talk about anything like an "Anthropocene research" as long as the agonistic approaches study completely different things that they either all happen to describe by the *word* Anthropocene or explicitly aim at replacing the Anthropocene with another concept.

If agonism about the object of study would qualify as a good thing, then an infinite number of objects of study could be called Anthropocene in a situation of contestation. Even if some approaches would be unable to communicate with each other by virtue of lacking a common object of study, shared concepts, vocabulary, concerns, imperatives, motifs, and so forth, all of them would be part of the larger family called "Anthropocene research." Humanities and social science scholars would probably recourse to a notion to capture that which holds "Anthropocene research" together to the extent that it can be at least named. The notion could describe a loose coherence that enables fluidity in the connectivities of belonging and in the manifold relationships between multiple elements in "Anthropocene Research." On this account, sharing an aspect with anything that is already considered as being part of "Anthropocene research" would qualify as part of the larger unit, all this leading to infinite regress. If practically anything could qualify as Anthropocene research merely by virtue of being able to share some of the concerns, motifs, and imperatives of some of the existing approaches, then new additions to the larger family—potentially bringing new concerns and motifs to which further approaches could connect—would constantly broaden this Anthropocene research, without any practical confines. Ultimately, a cacophony of views—some of them without sharing anything with some others—would pretend to argue with each other over the constitution of something that without any apparent reason they would all call "Anthropocene research."

Let's call this scenario *runaway agonism* in an "Anthropocene research" without boundaries.[36] The actual condition of the Anthropocene debate today looks closer to this than to the ideal of a productive agonistic pluralism advocated by Toivanen and his co-authors. For any meaningful transdisciplinary work, runaway agonism must be overcome. This is not to say that we need to pick one particular understanding of the notion of the Anthropocene and elevate it to being a shared one. Nor do I wish to pretend to be able to solve the conceptual confusion surrounding the notion of the Anthropocene. The point I wish to make is that any fruitful conception of transdisciplinary work needs *connective concepts* instead—concepts that bridge the respective concerns and agendas of participating knowledge formations.

By connective concepts in Anthropocene research I mean concepts that resonate with existing knowledge formations across the disciplinary landscape. They have the *potential* to prove to be useful for both humanities and social scientific research on the one hand and scientific research on the other. True enough, whether the potential can or cannot be realized will be assessed only after the fact, only after the work of concept creation. But normativity is an inherent feature of connective concepts. They point onward to a future state of research: they aim at shepherding existing knowledge formations toward the establishment of a new knowledge regime attuned to the investigation of the entangled human-technological-natural world as a new object of knowledge.

I think that Anthropocene research needs more such connective concepts. However, so long as the master concept meant to name the object of knowledge is not among such connective concepts, so long as it is the subject on which disciplines exercise their runaway antagonism, meaningful transdisciplinary work on a larger scale amounts to nothing more than a mirage.

Notes

1. Jan Zalasiewicz et al., "The Working Group on the Anthropocene: Summary of Evidence and Interim Recommendations," *Anthropocene* 19 (2017): 58.
2. Dolly Jørgensen, "Not by Human Hands: Five Technological Tenets for Environmental History in the Anthropocene," *Environment and History* 20 (2014): 480.
3. See also Colin N. Waters et al., "Can Nuclear Weapons Fallout Mark the Beginning of the Anthropocene Epoch?" *Bulletin of the Atomic Scientists* 71, no. 3 (2015): 46–57.
4. Here is the full voting result: "aluminium, 0; plastic, 3; fuel ash particles, 2; carbon dioxide concentration, 3; methane concentration, 0; carbon isotope change, 2; oxygen isotope change, 0; radiocarbon bomb spike, 4; plutonium fallout, 10; nitrate concentration/$\delta^{15}N$, 0; biostratigraphic extinction/assemblage change, 0; other (lead, persistent organic pollutants, technofossils), 3; uncertain, 2; abstain, 6." Zalasiewicz et al., "The Working Group on the Anthropocene," 58.
5. Zalasiewicz et al., "The Working Group on the Anthropocene," 57.
6. Will Steffen et al. "The Trajectory of the Anthropocene: The Great Acceleration," *The Anthropocene Review* 2, no. 1 (2015): 81–98.
7. Alexander Koch et al., "Earth System Impacts of the European Arrival and Great Dying in the Americas after 1492," *Quaternary Science Reviews* 207 (2019): 13–36.
8. Eric Schatzberg, *Technology: Critical History of a Concept* (Chicago: University of Chicago Press, 2018), 15.
9. Hannah Arendt, *Between Past and Future: Six Exercises in Political Thought* (New York: The Viking Press, 1961). 60.
10. Alf Hornborg, *The Power of the Machine: Global Inequalities of Economy, Technology, and Environment* (Walnut Creek: Altamira Press, 2001).
11. Alf Hornborg, "The Political Ecology of the Technocene: Uncovering Ecologically Unequal Exchange in the World-System," in *The Anthropocene and the Global Environmental Crisis: Rethinking Modernity in a New*

Epoch, eds. Clive Hamilton, François Gemenne and Christophe Bonneuil (London and New York: Routledge, 2015), 57–69.
12. Haff, P. K. "Technology and Human Purpose; The Problem of Solid Transport on the Earth's Surface," *Earth System Dynamics* 3 (2012): 149–156; Peter Haff, "Humans and Technology in the Anthropocene: Six Rules," *The Anthropocene Review* 1, no. 2 (2014): 126–136.
13. Haff, "Humans and Technology in the Anthropocene," 127.
14. Haff, "Humans and Technology in the Anthropocene," 127.
15. Haff, "Humans and Technology in the Anthropocene," 127.
16. Jan Zalasiewicz et al., "The Technofossil Record of Humans," *The Anthropocene Review* 1, no. 1 (2014): 40.
17. Eva Lövbrand et al., "Who Speaks for the Future of Earth? How Critical Social Science Can Extend the Conversation on the Anthropocene," *Global Environmental Change* 32 (2015): 212.
18. Robin, Libby. "Environmental Humanities and Climate Change: Understanding Humans Geologically and Other Life Forms Ethically," *WIREs Climate Change* 9 (2018): e499.
19. See, for instance, Rosi Braidotti, *The Posthuman.* Cambridge: Polity, 2013; Donna J. Haraway, *Staying with the Trouble: Making Kin in the Chthulucene* (Durham: Duke University Press, 2016), 30–57; Claire Colebrook, "What is the Anthropo-Political?," in Tom Cohen, Claire Colebrook and J. Hillis Miller, *Twilight of the Anthropocene Idols* (London: Open Humanities Press, 2016), 81–125; Stacy Alaimo, *Exposed: Environmental Politics and Pleasures in Posthuman Times.* Minneapolis: University of Minnesota Press, 2016.
20. Andreas Malm and Alf Hornborg, "The Geology of Mankind? A Critique of the Anthropocene Narrative," *The Anthropocene Review* 1, no. 1 (2014): 64.
21. Ian Angus, *Facing the Anthropocene: Fossil Capitalism and the Crisis of the Earth System* (New York: Monthly Review Press, 2016), 224–232.
22. Dipesh Chakrabarty, "The Politics of Climate Change Is More Than the Politics of Capitalism," *Theory, Culture & Society* 34, no. 2–3 (2017): 28; Julia Adeney Thomas, "Confronting Climate Change: The Uneasy Alliance of Scientists and Nonscientists in a Neoliberal World," *Environmental History* 23, no. 1 (2018): 177.
23. Julia Adeney Thomas and Michael A. Ellis, "Why 'Anthropocene' is Not the Same as 'Climate Change' and Why It Matters for Asia," *Global Sustainability*, forthcoming.
24. For more details see Zoltán Boldizsár Simon, "(The Impossibility of) Acting upon a Story That We Can Believe," *Rethinking History* 22, no. 1 (2018): 105–125; Zoltán Boldizsár Simon, "The Limits of Anthropocene

Narratives," *European Journal of Social Theory*, online first article (2018), DOI: https://doi.org/10.1177/1368431018799256

25. Tom Cohen and Claire Colebrook, "Vortices: On 'Critical Climate Change' as a Project," *South Atlantic Quarterly* 116, no. 1 (2017): 135.
26. For the positions in the internal humanities and social scientific debate against which the above arguments are directed, see Bruno Latour, "Why Has Critique Run out of Steam? From Matters of Fact to Matters of Concern," *Critical Inquiry* 30, no. 2 (2004): 225–248; Rita Felski, *The Limits of Critique* (Chicago: University of Chicago Press, 2015).
27. Christophe Bonneuil and Jean-Baptiste Fressoz, *The Shock of the Anthropocene: The Earth, History and Us*, trans. David Fernbach (London: Verso, 2016), 80.
28. Julia Adeney Thomas, "The Shock of the Anthropocene: The Earth, History, and Us," *Social History*, 42, no. 3 (2017): 458–460; Thomas, "Confronting Climate Change," 176–177.
29. Zalasiewicz et al., "The Working Group on the Anthropocene," 56.
30. Jussi Parikka, *The Anthrobscene* (Minneapolis: University of Minnesota Press, 2014).
31. For a discussion of both the Capitalocene and the Plantationocene, see Donna Haraway et al., "Anthropologists Are Talking – About the Anthropocene," *Ethnos* 81, no. 3 (2016): 535–564.
32. Angus, *Facing the Anthropocene*, 230–231.
33. T. Toivanen et al., "The Many Anthropocenes: A Transdisciplinary Challenge for the Anthropocene Research," *The Anthropocene Review* 4, no. 3 (2017): 183–198.
34. Toivanen et al., "The Many Anthropocenes," 186.
35. For the contrast between agonism and antagonism, see Chantal Mouffe, *On the Political* (London and New York: Routledge, 2005).
36. Runaway agonism is not antagonism in the sense that it does not necessarily entail that the views that share nothing with each other consider each other as enemies. However, when the respective views challenge the very foundations and premises of "Anthropocene research," runaway agonism may indeed resemble to an antagonistic relation.

Bibliography

Alaimo, Stacy. *Exposed: Environmental Politics and Pleasures in Posthuman Times*. Minneapolis: University of Minnesota Press, 2016.

Angus, Ian. *Facing the Anthropocene: Fossil Capitalism and the Crisis of the Earth System*. New York: Monthly Review Press, 2016.

Arendt, Hannah. *Between Past and Future: Six Exercises in Political Thought.* New York: The Viking Press, 1961.

Bonneuil, Christophe, and Jean-Baptiste Fressoz. *The Shock of the Anthropocene: The Earth, History and Us.* Translated by David Fernbach. London: Verso, 2016.

Braidotti, Rosi. *The Posthuman.* Cambridge: Polity, 2013.

Chakrabarty, Dipesh. "The Politics of Climate Change Is More Than the Politics of Capitalism." *Theory, Culture & Society* 34, no. 2–3 (2017): 25–37.

Cohen, Tom, and Claire Colebrook. "Vortices: On 'Critical Climate Change' as a Project." *South Atlantic Quarterly* 116, no. 1 (2017): 129–143.

Colebrook, Claire. "What is the Anthropo-Political?," In Tom Cohen, Claire Colebrook and J. Hillis Miller, *Twilight of the Anthropocene Idols*, 81–125. London: Open Humanities Press, 2016.

Felski, Rita. *The Limits of Critique.* Chicago: University of Chicago Press, 2015.

Haff, P. K. "Technology and Human Purpose; The Problem of Solid Transport on the Earth's Surface." *Earth System Dynamics* 3 (2012): 149–156.

Haff, Peter. "Humans and Technology in the Anthropocene: Six Rules." *The Anthropocene Review* 1, no. 2 (2014): 126–136.

Haraway, Donna J. *Staying with the Trouble: Making Kin in the Chthulucene.* Durham: Duke University Press, 2016.

Haraway, Donna et al. "Anthropologists Are Talking – About the Anthropocene." *Ethnos* 81, no. 3 (2016): 535–564.

Hornborg, Alf. *The Power of the Machine: Global Inequalities of Economy, Technology, and Environment.* Walnut Creek: Altamira Press, 2001.

Hornborg, Alf. "The Political Ecology of the Technocene: Uncovering Ecologically Unequal Exchange in the World-System." In *The Anthropocene and the Global Environmental Crisis: Rethinking Modernity in a New Epoch*, edited by Clive Hamilton, François Gemenne and Christophe Bonneuil, 57–69. London and New York: Routledge, 2015.

Jørgensen, Dolly. "Not by Human Hands: Five Technological Tenets for Environmental History in the Anthropocene." *Environment and History* 20 (2014): 479–489.

Koch, Alexander et al. "Earth System Impacts of the European Arrival and Great Dying in the Americas after 1492." *Quaternary Science Reviews* 207 (2019): 13–36.

Latour, Bruno. "Why Has Critique Run out of Steam? From Matters of Fact to Matters of Concern." *Critical Inquiry* 30, no. 2 (2004): 225–248.

Lövbrand, Eva et al. "Who Speaks for the Future of Earth? How Critical Social Science Can Extend the Conversation on the Anthropocene." *Global Environmental Change* 32 (2015): 211–218.

Malm, Andreas, and Alf Hornborg. "The Geology of Mankind? A Critique of the Anthropocene Narrative." *The Anthropocene Review* 1, no. 1 (2014): 62–69.

Mouffe, Chantal. *On the Political.* London and New York: Routledge, 2005.

Parikka, Jussi. *The Anthrobscene*. Minneapolis: University of Minnesota Press, 2014.
Robin, Libby. "Environmental Humanities and Climate Change: Understanding Humans Geologically and Other Life Forms Ethically." *WIREs Climate Change* 9 (2018): e499.
Schatzberg, Eric. *Technology: Critical History of a Concept*. Chicago: University of Chicago Press, 2018.
Simon, Zoltán Boldizsár. "(The Impossibility of) Acting upon a Story That We Can Believe." *Rethinking History* 22, no. 1 (2018): 105–125.
Simon, Zoltán Boldizsár. "The Limits of Anthropocene Narratives." *European Journal of Social Theory*, online first article (2018). https://doi.org/10.1177/1368431018799256
Steffen, Will et al. "The Trajectory of the Anthropocene: The Great Acceleration." *The Anthropocene Review* 2, no. 1 (2015): 81–98.
Thomas, Julia Adeney. "The Shock of the Anthropocene: The Earth, History, and Us." *Social History* 42, no. 3 (2017): 458–460.
Thomas, Julia Adeney. "Confronting Climate Change: The Uneasy Alliance of Scientists and Nonscientists in a Neoliberal World." *Environmental History* 23, no. 1 (2018): 172–182.
Thomas, Julia Adeney, and Michael A. Ellis. "Why 'Anthropocene' is Not the Same as 'Climate Change' and Why It Matters for Asia." *Global Sustainability*, forthcoming.
Toivanen, T. et al. "The Many Anthropocenes: A Transdisciplinary Challenge for the Anthropocene Research." *The Anthropocene Review* 4, no. 3 (2017): 183–198.
Waters, Colin N. et al. "Can Nuclear Weapons Fallout Mark the Beginning of the Anthropocene Epoch?" *Bulletin of the Atomic Scientists* 71, no. 3 (2015): 46–57.
Zalasiewicz, Jan et al. "The Technofossil Record of Humans." *The Anthropocene Review* 1, no. 1 (2014): 34–43.
Zalasiewicz, Jan et al. "The Working Group on the Anthropocene: Summary of Evidence and Interim Recommendations." *Anthropocene* 19 (2017): 55–60.

CHAPTER 4

Epochal Thinking and Anthropogenic Catastrophe

Abstract A new kind of epochal thinking is emerging against the backdrop of the perception that human capacities have increased to such an extent that humanity has become a threat to the planet and to its own existence, developing a capacity even for self-extinction. In reviewing prospects of anthropogenic planetary catastrophe, this chapter outlines how the new epochality took root in post-Second World War societies. It introduces three instances that exemplify the new epochal: the ongoing "sixth extinction" event, the potential transgression of "planetary boundaries," and the prospect of a "technological singularity." In the course of discussion, the chapter touches upon several key themes such as anthropocentrism, the future as threat, the potential entry into a "new reality," and the boundaries of the notion of the Anthropocene.

Keywords Epochal thinking • Human capacities • Anthropocentrism • Sixth extinction • Planetary boundaries • Technological singularity • New reality

Capacity for Self-Extinction

How did the world of human affairs, the technological domain, and the natural world end up so inextricably bound? And what exactly drives this entanglement? Answering these questions is of utmost importance for

Z. B. Simon, *The Epochal Event*, Palgrave Studies in the History of Science and Technology,
https://doi.org/10.1007/978-3-030-47805-6_4

any effort that aims at understanding our current condition. In light of the larger discussion on the Anthropocene, there is an extent to which the answer is not particularly surprising: human activity lurks behind it of course. Yet this does not say much; it only identifies a source. The intriguing aspect concerns the peculiar character of human activity behind the human-technology-nature entanglement, namely, its assumed capacity to launch catastrophic and self-destructive events on a planetary scale.

The challenges Western societies lately conceive of as most pressing, derive from the perception that the human being has become the greatest threat both to its own survival and to the flourishing of other life-forms in an entangled human-technological-natural world. As to the latter, there is typically little confusion about what the extinction of species means (be it human-induced or not). The case is more difficult with human self-extinction, which is not necessarily viewed as the eradication of biological life and the dying out of the human species.

Claire Colebrook's distinction between three senses of extinction perfectly speaks to this point. According to Colebrook, there is "the now widely discussed sixth great extinction event (which we can begin to imagine, witnessed by humans even if in anticipation)"; then there is "extinction by humans of other species (with the endangered species of the 'red list' evidencing our destructive power)": and, finally, we also have "self-extinction, or the capacity for us to destroy what makes us human."[1] As will be soon apparent, the first two senses are essentially the same because the sixth extinction is not something we only imagine and anticipate but one that is ongoing and that science tells us we cause. The point I wish to make, however, is about Colebrook's third sense of extinction.

In Colebrook's classification, human self-extinction refers to the annihilation of certain attributes that appear as specifically human and not to the discontinuation of the existence of biologically defined humans. It seems that human self-extinction can mean extinction in a cultural sense; only the extinction of other species must be biological. Let's bracket the fact that reserving the privilege for humans to go extinct merely by virtue of losing certain attributes may easily come out as human exceptionalism even according to Colebrook's own critical humanities sensibility. More important to note at this point is that such non-literal sense of human self-extinction also underpins debates on human enhancement and on transhumanist prospects. In this context, expressing anxiety about the prospect of enhancement resulting in the loss of certain qualities, virtues, and subjectivities previously attributed to being human is a position usually associated with bioconservativism.[2]

Conceptions of non-biological human self-extinction aside, this chapter will focus on the ways in which the human being has become both the source and the subject of extinction in the most literal sense. In doing so, it will attempt to recount the emergence of a new kind of epochal imagination fueled by the perception of human beings acquiring unprecedented powers to the extent that we are now threatened with a self-authored human catastrophe on a planetary scale. The rise of such epochal thought is the condition of possibility and the broader context of epochal events (yet to be conceptualized properly by Chapter 6). As they emerge jointly and as the events themselves are the clearest indicators of the new epochal imagination, the chapter will pay special attention to the three most frequently discussed instances of the new epochal. The events in question are the ongoing sixth extinction, the prospect of a technological singularity, and the potential transgression of planetary boundaries.

Two of these events are closely related to the broadly construed Anthropocene debate. Whereas the sixth extinction of species in Earth history is usually discussed in the biological approach to the Anthropocene (according to the fourfold categorization of Toivanen et al. mentioned in the previous chapter),[3] the concept of planetary boundaries is an Earth system science attempt to identify a zone of safety, an operating space within certain environmental limits, known to support human life on the planet. The notion of technological singularity is rarely mentioned together with the sixth extinction and the potential transgression of planetary boundaries. Intending to capture the point of the creation of greater-than-human intelligence in terms of a singular event beyond which the future is humanly unfathomable, the notion features mainly in discussions on artificial intelligence (AI)—meaning strong AI, that is, artificial general intelligence—and human enhancement.[4]

In the coming pages, I will organize the discussion of the instances of the new epochal and the emergence of a new epochal thinking around three themes. First, I will explore the relationship between epochal transformations and anthropocentrism; second, I will say a few words about the perception of the future in terms of threats; and third, I will discuss the future as an entry into a "new reality." Finally, I will end the chapter with a discussion of how the new epochal reactivates older patterns of epochal thinking.

Anthropocentrism

Human activity is increasingly identified today as the drive behind rapid biodiversity loss, even as the primary agent of an ongoing sixth mass extinction event in Earth history. Out of the previous five—the so-called Big Five—the fifth extinction is fairly well known. About 66 million years ago, it resulted in the extinction of dinosaurs.

In an extinction event, the extinction rate of species is significantly higher than "natural" or "background" extinction rates. The estimation of current human-induced extinction rates of species varies between 100 and 1000 times higher than those "background" rates, and estimations of future extinction rates reach the figure of 10,000 times higher.[5] As to the question of the ongoing extinction event being human-induced, such is the consensus of more than 15,000 scientists who signed the "World Scientists' Warning to Humanity: A Second Notice" in 2017.[6] The first warning was issued by the Union of Concerned Scientists in 1992, and supported by 104 Nobel laureates among the more than 1700 scientists endorsing it.[7] The first warning has already set the tone and upped the stakes with the following opening paragraph:

> Human beings and the natural world are on a collision course. Human activities inflict harsh and often irreversible damage on the environment and on critical resources. If not checked, many of our current practices put at serious risk the future that we wish for human society and the plant and animal kingdoms, and may so alter the living world that it will be unable to sustain life in the manner that we know. Fundamental changes are urgent if we are to avoid the collision our present course will bring about.[8]

Although the first warning did not yet name a "sixth extinction" as the second warning does, among its many concerns—ranging from water resources and deforestation to population growth—it explicitly referred to the rates of species loss that has later been commonly associated with an ongoing sixth extinction event. It claimed that "the irreversible loss of species, which by 2100 may reach one third of all species now living, is especially serious" and cautioned that "our tampering with the world's interdependent web of life" may lead to further effects, "including unpredictable collapses of critical biological systems whose interactions and dynamics we only imperfectly understand."[9]

The imperfect understanding of the consequences of human activity triggering planetary changes was accompanied by two things of which the warning of 1992 was perfectly unambiguous. First, it made clear that the human and the natural worlds are intertwined far more substantially than in the sense of a mere cause-and-effect relationship in which the former transforms the latter. By asserting that "the greatest peril is to become trapped in spirals of environmental decline, poverty, and unrest, leading to social, economic and environmental collapse,"[10] it called attention to the fact that the human and the natural worlds may fall together. Second, the warning underscored the extent to which an anthropocentric view and a universalistic attitude are inescapable if one takes the situation seriously. It claimed that acting on concerns for environmental damage is "enlightened self-interest," because "we all have but one lifeboat," because "no nation can escape from conflicts over increasingly scarce resources," and because the mass migrations likely to be caused by "environmental and economic instabilities" will have "consequences for developed and undeveloped nations alike."[11]

As important as these words are in raising awareness of an ongoing anthropogenic crisis, they may sound rather contentious in other contexts. Humanities criticism would likely tear them apart by questioning if it really is "human" self-interest and not only the self-interest of the rich. Critics would argue that previous experience demonstrates that the rich pretty much tend to have their lifeboats in situations of crisis. Dipesh Chakrabarty, whose views are congruent with the 1992 warning in suggesting that not having lifeboats is precisely what distinguishes the current crisis from previous crises of capitalism,[12] has already been fiercely criticized along these lines.[13] On the other hand, a rejoinder to such criticism could argue that we should not habitually reject new views out of hand just because they remind us of previously critiqued older perspectives. It may very well be that the crisis we face today is precisely the one that defies previous experiences.

Another line of criticism, associated with emancipatory humanities scholarship and the aim of overcoming anthropocentrism, would likely take issue with the "self-interest" component of the 1992 warning. It would argue that one may actually have non-anthropocentric environmental concerns for the nonhuman and more-than-human world.[14] A response to this criticism would point out that even those who do not think that human self-interest motivates actions to avoid planetary catastrophe cannot escape the fact that their anti-anthropocentric agenda necessarily remains *their* human agenda. But this is neither the only argument for an inescapable anthropocentrism nor the most profound one. Anthropocentrism in a philosophical and phenomenological sense is not even the target of anti-anthropocentric humanities. Understanding the perplexities surrounding the human element in the overall predicament requires further questioning concerning the motive of human self-interest.

To begin with, the warning invokes self-interest in order to facilitate action. And when it comes to action, anthropocentrism—in a sense that extends beyond the philosophical anthropocentrism entailed by the contention that non-anthropocentrism is a human agenda—indeed seems to be inescapable. The reason for this lies in the planetary character of the required action. If we aim at acting in order to avoid planetary catastrophe, then the sheer fact of being able to act (or the sheer fact of merely thinking that we are able to act) on such a scale implies the workings of tremendous human powers. This is the sense in which even critiques of anthropocentrism and human exceptionalism cannot avoid invoking the very anthropocentrism they criticize.

Recommendations to overcome anthropocentrism through acing upon the recognition of the wrongdoings of human exceptionalism simply cannot escape relying on the workings of anthropocentric powers. As an example, consider Eileen Crist's views, appearing in *Science* in 2018, and laudably bringing a wide exposure of humanities arguments to a dominantly scientific or science-oriented audience. Crist's point of departure is the mass extinction of species and planetary-scale damages. She attributes them to a worldview affiliated with attitudes she describes by the terms "human expansionism," "human supremacy," "entitlement," and "superiority," meaning that "this worldview esteems the human as distinguished entity that is superior to all other life forms and is entitled to use them and the places they live."[15] It is due to this worldview—spreading from the Western world over the globe according to Crist—that looking at nonhumans and the ecosphere means that we see only resources to exploit without being able to see the costs and the violence involved. Furthermore, it appears to Crist that in this human-centered worldview we cannot even put a limit to our wrongdoings, as "embracing limitation" would come out as "unbefitting of human distinction."[16] Nevertheless, limitation is precisely what she suggests: to "scale down" and "pull back."

Crist's claim that the worldview of human mastery "hinders the recognition that scaling down and pulling back is the most farsighted path forward"[17] clearly contradicts the appeal to an "enlightened self-interest" in the World Scientists' Warning to Humanity issued in 1992.[18] What the latter says is precisely that we have reached a point where our very human-centered views require us to pull back. I tend to think that the warning of 1992 makes a good point. And its recommendations based on human self-interest do not differ much from those of Crist's based on an anti-anthropocentric agenda. The similarities between the two are, in fact, haunting.

The warning's claim that "we must give high priority to efficient use of energy, water, and other materials, including expansion of conservation and recycling"[19] is echoed by Crist who contends that "scaling down involves reducing the overall amount of food, water, energy, and materials that humanity consumes and making certain shifts in what food, energy and materials are used."[20] As to what Crist calls "pulling back our presence from the natural world," she thinks that "achieving continental-scale protection of terrestrial and marine habitats will enable sharing the Earth generously with all its life forms."[21] And again, the warning aims at achieving the same goals. It claims that "we must bring environmentally damaging activities under control to restore and protect the integrity of the earth's systems we depend on," by which it means not only that we need to "move away from fossil fuels" but also that "we must halt deforestation, injury to and loss of agricultural land, and the loss of plants, animals, and marine species."[22]

In order to achieve all this, the warning thinks that "a new ethic is required – a new attitude towards discharging our responsibility for caring for ourselves and for the earth."[23] Crist is of course much more explicit about this as she already advocates the anti-anthropocentric ethics developed by humanities scholarship in the last two decades. She ends her essay by stating the agenda according to which pursuing "scaling down and pulling back the human factor requires us to reimagine the human in a register that no longer identifies human greatness with dominance within the ecosphere and domination over nonhumans."[24]

I do not wish to argue here neither for nor against the desirability of any kind of ethics that may be required to manage the situation. What I want to point out is only that even the anti-anthropocentric ethics advocated by Crist remains human-centered inasmuch as it implies that the fate of the planet hinges upon human actions and human capacities to act. Not just any action but planetary-scale action, and not just any human capacities but unparalleled and historically unprecedented ones. Even if one denies that the motivation for planetary-scale human action is human self-interest, the action remains planetary-scale human action with the assumed potential to fix issues that, again, matter for us. We are the ones taking issue with our own perceived "domination," and we are the ones concerned with sharing the Earth "generously" and fixing the damages we have caused. Acting on the recognition of anthropogenic transformations on a planetary scale simply cannot occur without anthropocentrism in an elementary sense, without ideas of human mastery, and without having a recourse to human capacities perceived as exceptional, even if we know that this is precisely what led to the current crisis. We need to figure out whether the unprecedented human capacities are determined to be used the way they have been used or not, and whether the consequences of using those capacities are inevitable or not.

In any case, this is no easy situation: wanting to act to stop or even to decrease the rates of a human-induced mass extinction of species necessitates the very same powers that we now believe to have led to catastrophe. What happens when we recourse to such powers (even if we eventually take Crist's view and scale down and pull back)? At the end of her book *The Sixth Extinction: An Unnatural History*, Elizabeth Kolbert mentions two scenarios with respect to coping with extinction. In the first one, "human ingenuity will outrun any disaster human ingenuity sets in motion."[25] The other is human extinction due to human activity. Now, these are obviously extreme projections, and there are many other plots one can imagine in between the two poles of continuing the project of human mastery over nature for the better on one side and human self-extinction on the other. What I wish to point out is only that as soon as one takes seriously the warnings concerning anthropogenic planetary changes, there is an extent to which anthropocentrism is inescapable.[26]

This is not only a matter of bringing about immense changes but also a matter of coping with them, steering them in one way or another, and even acting in order to avoid them. Action on a planetary scale necessarily assumes the existence of capacities for an overall planetary management that oscillates between the poles of destruction and creation, and between the poles of scaling up and scaling down. Instead of advocating singular agendas as potential solutions, perhaps we need to learn to inhabit the situation by navigating carefully among conflicting imperatives. But more on this in the last chapter. For now, let's turn to the question of why we conceive of our current situation in terms of threats in the first place.

The Future as Threat

The perception that human beings acquired previously unimaginable capacities with destructive and self-destructive potential is even more omnipresent in debates over new technologies. Biotechnology, human enhancement technologies, AI research, nanotechnology, and so on are increasingly discussed in terms of risks at best and in terms of their potential to end humanity in one way or another (either by leaving behind a human condition or by eradicating life on Earth) at worst.[27] This is not to say that biotechnology or genome editing is not considered of great promise in the medical context or that AI research is not perceived as yielding beneficial results. Nor is it to say that human capacities have never been discussed before in terms of a self-harming potential. It is only to say that there is a novelty in technologies testifying unparalleled human powers, namely, their perceived potential to result in large-scale societal collapse or even human self-extinction.

Let's consider an example that remains in the realm of warnings by scientists. Astrophysicist and cosmologist Martin Rees opens his 2003 book (which features the phrase "A Scientist's Warning" in its long subtitle) with the following paragraph:

> Science is advancing faster than ever, and on a broader front: bio-, cyber- and nanotechnology all offer exhilarating prospects; so does the exploration of space. But there is a dark side: new science can have unintended consequences; it empowers individuals to perpetrate acts of megaterror; even innocent errors could be catastrophic. The "downside" from twenty-first century technology could be graver and more intractable than the threat of nuclear devastation that we have faced for decades. And human-induced pressures on the global environment may engender higher risks than the age-old hazards of earthquakes, eruptions, and asteroid impacts.[28]

Rees compares the potential consequences of anthropogenic environmental changes to the threat of natural disasters. But the risks associated with latest technologies compare only to another human-made technology-related threat: the prospect of a nuclear holocaust. Indeed, technology has already become the greatest source of anxiety as the means of potential human self-destruction with the development of nuclear weaponry. Still within genre confines, the most illustrative act in this respect is the 1947 introduction of the symbolic Doomsday Clock by the *Bulletin of the Atomic Scientists.* The intention behind setting the Clock with a countdown to midnight (the time of the potential doomsday) is to convey the sense of how close humanity has come to a self-authored existential catastrophe. Today, the considerations of the Board of the *Bulletin* in setting the Clock include not only the nuclear threat but also the anthropogenic risks of climate change and recent technologies, and the Clock shows a hundred seconds to midnight, the closest time ever.[29]

Pointing out that the future started to appear as a self-authored threat to humanity following the realization of a potential self-annihilation by nuclear weaponry does not mean that a sense of inevitable doom unquestionably dominated post-Second World War expectations of the future. As Jenny Andersson brilliantly explores in her recent book *The Future of the World*,[30] the postwar period brought about manifold ways in which the future became an object of knowledge production. The emerging postwar futures constituted a site of struggle over how to shape the world. Yet, diverging perceptions and scenarios aside, the future as a problem emerged on the premise that "the post-war world was, more than any previous historical world, marked by the idea of human influence, and with the idea of unprecedented influence came new conceptions of consequence, reach, and responsibility."[31] The realization of the nuclear threat as such unprecedented human influence of potentially apocalyptic consequences effected even utopian thought. According to Andersson, already in the late 1960s (before the rise of futurology turned the tide toward more conscious technocratic efforts to engineer the future), "the only possible site of utopian energy" was "the idea of human beings themselves, as the harbingers of the apocalypse, but also as the sole saviours of the world."[32]

The ambivalence in expectations of the future Andersson detects as emerging in the early postwar decades remained the default setting even today. It equally informs discussions on anthropogenic changes in the Earth system and recent debates over the prospects of biotechnology, human enhancement, and the potential development of superintelligence. Ambivalence does not of course mean a 50–50 chance of things going either good or bad in the future. Rather, it means that although the dominant context of the Anthropocene debate is that of a potential catastrophe, an ecomodernist pole advocates the idea that modern technologies "offer a real chance of reducing the totality of human impacts on the biosphere," and, consequently, embracing "these technologies is to find paths to a good Anthropocene."[33] Similarly, in the debate on the perils of recent technological prospects of biotechnology and human enhancement, a transhumanist pole advocates the idea of making use of those technologies as means of escaping the biological limitations of our human constitution against the more pessimistic views of a bioconservative pole.[34] If the destructive poles nevertheless look dominant in both public and academic debates, it is because the very creative powers of promise reached a tipping point. Since the development of nuclear weaponry,[35] technology as the very source of a promise of making the world a better place (a premise informing Western modernity) has increasingly become perceived also as the primary means of destruction and self-authored human catastrophe in its structural workings.[36]

New Reality

The specter of self-authored threats to humanity is, I believe, the most apparent symptom of a larger phenomenon: the growing perception that epochal changes are about to take place. Regardless of whether the expectation of such changes concerns apocalyptic ways of human self-destruction or desired futures in an engineered fashion, the prospects are about human-induced epochal transformations. Not merely historical-epochal changes in the world of human affairs and not simply geological-epochal changes as we know them from the periodization of Earth history, but epochal transformations in the entangled human-technological-natural world.

The threat of the complete unknown and the perceived unprecedented unleash a new kind of epochal imagination that cannot even see beyond the immense changes it envisions. This is the ground on which technological powers are discussed today as having the potential to kick off a new era of posthumanity, inhabited by beings other than human. The new era is expected to come about either as a result of runaway human enhancement, or as a result of the creation of artificial superintelligence. The notion that describes the transition is technological singularity,[37] a point of no return. As soon as greater-than-human intelligence is created, whatever happens afterwards belongs to a posthuman realm, inaccessible to the human mind. In the words of Vernor Vinge—who popularized the concept at a NASA conference in 1993—the technological singularity

> is a point where our models must be discarded and a new reality rules. As we move closer and closer to this point, it will loom vaster and vaster over

> human affairs till the notion becomes a commonplace. Yet when it finally happens it may still be a great surprise and a greater unknown.[38]

Even if Vinge does not explicitly refer to a new epoch and does not even use the word "epoch," it can hardly be denied that a transformative event that signals an entry into a "new reality" with contours unknown to pre-evental thought is what we tend to associate with a change that opens a radically new era. Chapter 6 will explore in more details how radical and extreme such an epochal change that opens up a "new reality" may be. For now, what matters is simply the epochal character of the expected change.

The epochal imagination underlying approaches to runaway anthropogenic changes in the Earth system is of the same structure. In this context, the concept of "planetary boundaries" is the most instructive, referring to the environmental limits within which humanity can safely operate as relative to the functioning of the Earth system. Once those boundaries are crossed, abrupt and catastrophic changes may follow, with humanity being unable to further foresee that which it launched upon itself.

The original paper that introduced the notion in 2009 identified nine planetary boundaries (including one concerning biodiversity loss), while the updated paper published in 2015 further elaborated on the approach.[39] The complex interactions among the processes covered by the proposed boundaries—such as ocean acidification, land-system change, genetic diversity, or climate change—are beyond the scope of studies at the present stage. But even without profound and tremendously complex investigations concerning the question of how various planetary processes might interact, trigger, and potentially strengthen abrupt changes, researchers of the planetary boundary framework already claim with much certainty that a continuing trajectory can be catastrophic. "With an uncomfortably high probability," it leads away from Holocene conditions—"the only state of the planet that we know for certain can support contemporary human societies"—"to a very different state of the Earth system, one that is likely to be much less hospitable to the development of human societies."[40]

The New Epochal

The sixth extinction, the prospect of a technological singularity, and the potential transgression of planetary boundaries are the main representatives of the rise of an epochal thought that was inconceivable before postwar times and could emerge only due to the growing perception of unprecedented human powers. They are the most powerful instances of a new kind of epochal event and a new notion of epochal transformation that conceives of change as the opening up of a "new reality"—a reality completely unfathomable prior to the event that opens it up.

The emergence of a new kind of epochal thinking is, however, only one side of the coin. The other side is the impulse this provides to revisiting ideas about older kinds of epochal transitions in several related and non-related contexts. As an example that somehow blends old and new, one can think of the aforementioned trend in the humanities and the social sciences to develop an anti- or non-anthropocentric stance. Although anti-anthropocentric approaches also work within the human-technology-nature entanglement, they oftentimes brand themselves as counter-reactions to the postwar view of humanity achieving exceptional capacities.

In the shape of critical posthumanism, anti-anthropocentric scholarship calls for the development of a worldview that places the human being within the web of planetary life.[41] Critical posthumanism may express a technophilic attitude, but it typically remains hostile toward technological visions that aim at bringing about a new epoch of technological posthumanity. Against the potential creation of technological posthuman *subjects* with exceptional and even greater-than-human capacities, critical posthumanism focuses on developing a new human post-anthropocentric *subjectivity* of biologically human beings that, rather confusingly, it also calls posthuman.[42]

Discourses on a technological posthumanity (with better-than-human posthuman subjects) and on critical posthumanism (with a new anti-anthropocentric subjectivity) equally advance epochal claims. Whereas the former concerns, as Francis Fukuyama fears, a posthuman stage in history through the realization of transhumanist aspirations, runaway enhancement, and biotechnologies,[43] the latter advocates a tectonic rearrangement of knowledge and the development of a posthuman humanities or a posthumanities in place of recent humanities knowledge production.[44] The former as a radical entry into a "new reality" of machine superintelligence and enhanced superhumans is an instance of the new epochal. Transforming the humanities into posthumanities, however, invokes a more familiar type of epochal change. As a historical-epochal claim that presents the latest development in ways of knowledge production, it remains committed to the principles of modern (Western) historical thought.

There is nothing surprising in witnessing such claims. It seems rather "natural" that new kinds of human-induced and extinction-related visions of epochal changes in the context of the human-technology-nature entanglement reactivate already available patterns of epochal thinking and bring them to the foreground. Historical-epochal and geological-epochal changes are comparably modest ways in which epochal thinking has been present both in human and in natural history throughout Western modernity. The paradigmatic example of a historical-epochal transformation is the fall of the *ancient régime* through the historical event of the French Revolution and the subsequent emergence of new political systems. On a more profound level, one can also think of how history departments are still structured (typically, although by now not exclusively) along historical-epochal divisions: ancient history, medieval history, early modern history, modern history, and contemporary history. As to geological-epochal claims, recall the discussion of Chapter 3, in which geology turned out to be the large-scale natural scientific counterpart of the discipline of history, recounting the history of the Earth in the structurally similar gradual way as historical studies recount human history. The handiest example of a traditional geological-epochal claim is of course the Anthropocene as a proposed (but not ratified) geological epoch, marking the era of the human-technology-nature entanglement.

To make things more complex, in a broader cultural understanding of the notion, the Anthropocene can also be conceived of as a historical-geological epoch *within* which new types of epochal thought emerge (the epochal imagination of the entangled human-technological-natural world), which, in turn, reactivates old types of epochal thinking (more familiar historical-epochal and geological-epochal claims). In fact, old and new types of epochal thinking merge today to an extent that in their particular manifestations it would be difficult to separate them. The best example to illustrate the reactivation of the old by the new and their blending is the recent publication of Buffon's *The Epochs of Nature*, an eighteenth-century work. It was translated into English from the French original for the first time in its entirety in 2018.[45] The translation and the editorial work was carried out by leading figures of the Anthropocene debate, who justify their endeavor as follows:

> Hence, some closer acquaintance with this savant of pre-Revolutionary France may be useful to a wide community, now that natural history and cosmology have reunited in planetary systems science, and in other

> developments such as Big History, and where many specialist sciences that have emerged in the intervening three centuries are now again working together. Not every polymath synthesizer of today can be assumed to be a reader of French, yet the ideas of Buffon take on a new significance as stratigraphers reconsider the idea of "epochs," consider anew how the present (and future) relate to the past—and ask the question whether the epoch of the Holocene is now completed, to be replaced by a yet newer epoch.[46]

Reconsidering familiar patterns of epochal thought in light of the new brings to the fore various themes related to historicity and temporality. To begin with, given that epochal thinking slices up time to periods which then serve as tools of historical comprehension, it comes as no surprise that the question of the onset of the Anthropocene as a potentially new epoch is one of the central issues of the larger debate.[47] But to get a fuller picture of how questions of historicity and temporality pervade the entire discussion, recall the voting of the Anthropocene Working Group discussed in Chapter 3. The six questions they tried to settle are the following: "is the Anthropocene stratigraphically real"; "should the Anthropocene be formalized"; "hierarchical level of the Anthropocene" (roughly speaking, the question of the type of the period in which a geological understanding of "period" is one among the options alongside "epoch," "era," etc.); "when should the Anthropocene begin"; in what way should it be defined; and "what is the best primary marker for the Anthropocene."[48] Most of these questions, if not all of them, are either directly or indirectly related to periodization and chronology as fundamental tenets of historical thinking. Little wonder that, coinciding with the Anthropocene debate, historians and theorists of historical thinking readdress these questions today, while the overall theme of historical time—the larger context in which questions of periodization and chronology are discussed with a renewed interest—has become an increasingly widely discussed topic in the new millennium.[49]

The reactivation of old kinds of epochal claims by novel kinds (and their blending) means that both kinds are equally pertinent today. Together they attest to a wider societal investment in expecting epochal changes to take place and perceiving them as already taking place today. It is tempting to link this phenomenon to an "epochal consciousness," a term that Dipesh Chakrabarty recently borrowed from Karl Jaspers in his discussion of the Anthropocene and anthropogenic climate change.[50] Although Jaspers coined the term in the interwar period (1931 in the German edition), Chakrabarty reads it together with Jaspers's later reflections on the nuclear threat in the early 1960s,[51] giving the term a broad range of meanings. Among them is one that Chakrabarty mentions only passingly, namely, that "epochal consciousness seeks to ingest a slab of historical time in its entirety."[52]

Capturing the *Zeitgeist* of an epoch, however, cannot be farther away from what I mean by the wider societal engagement in expecting epochal changes to take place and perceiving them as already taking place. What I would like to emphasize is not a vision of the essence of the latest epoch, but a growing societal sense that our time is the time of epochal transformations. Although this means that I have to resist the temptation to borrow Chakrabarty's borrowed term, I would nevertheless like to credit his work for calling attention to the centrality of a certain epochal awareness in twenty-first-century (Western) societies.

Next to the affiliation with Chakrabarty's work, I also want to make clear how the question of epochal change connects to my previous work on an emerging historical sensibility other than the one which has been associated with the processual and developmental idea of history in Western modernity.[53] The key point of connection is the notion of "unprecedented change," mentioned in Chapter 2. It attempts to adequately label a novel *type* or configuration of change that disconnects past and future. Unprecedented change does not merely mean change *in* the condition of a certain subject. Instead, it refers to the bringing about of a *new subject with no prior existence*, a subject whose coming to existence is not conceived of as merely unfolding from a previous state of affairs and past potential. On a larger scale, to characterize the overall sense of historicity that revolves around a series of unprecedented changes, I also introduced some preliminary thoughts concerning what I called an emerging "evental historical sensibility." The reason behind attributing an evental character to the new historical sensibility was that unprecedented change seemed to me to be the result of singular game-changer events, some of them mentioned above. Yet, in talking about unprecedented change, I did not have much to say neither about the conceptual outlook of such events nor about their context. I simply focused on the question of change.

It seems reasonable to align the notions of unprecedented change and evental historical sensibility with the expected epochal transformations in the human-technological-natural world. A new epochality is, I think, a key constituent of the equally novel historical sensibility. Transgressing "planetary boundaries," potentially arriving at a "technological singularity," and being in the midst of a sixth extinction can be considered as instances of both unprecedented change and the new kinds of events that signal epochal transformations. They are the paradigmatic events that revive more familiar patterns of epochal thought today, and they are the events that I intend to capture by the connective concept "epochal event" in the remaining pages of this book.

Notes

1. Clarie Colebrook, "Framing the End of Species," *symplokē* 21, no. 1–2 (2013): 51.
2. For bioconservative views see Francis Fukuyama, *Our Posthuman Future: Consequences of the Biotechnology Revolution* (New York: Farrar, Straus and Giroux, 2002); Michael J. Sandel, *The Case Against Perfection: Ethics in the Age of Genetic Engineering* (Cambridge, MA: The Belknap Press of Harvard University Press, 2007).
3. T. Toivanen et al., "The Many Anthropocenes: A Transdisciplinary Challenge for the Anthropocene Research," *The Anthropocene Review* 4, no. 3 (2017): 183–198.
4. The fact that the technological singularity is one of the most spectacular instances of the new epochal provides yet another good reason for not to confine the rise of epochal thinking to the most common understandings of the Anthropocene.
5. On the sixth extinction see both scientifically framed approaches and instances of popular science writing. Richard Leakey and Roger Lewin, *The Sixth Extinction: Patterns of Life and the Future of Humankind* (New York: Anchor Books, 1996); Stuart J. Pimm and Thomas M. Brooks, "The Sixth Extinction: How Large, Where, and When?," in *Nature and Human Society: The Quest for a Sustainable World*, eds. Peter H. Raven and Tania Williams (Washington, D.C.: National Academy Press, 2000), 46–62; Elizabeth Kolbert, *The Sixth Extinction: An Unnatural History* (London: Bloomsbury, 2014). Specifically on extinction rates see De Vos, Jurriaan M. et al., "Estimating the Normal Background Rate of Species Extinction," *Conservation Biology* 29, no. 2 (2014): 452–462.
6. William J. Ripple et al., "World Scientists' Warning to Humanity: A Second Notice," *BioScience* 67, no. 12 (2017): 1026–1028.
7. Union of Concerned Scientists, "World Scientists' Warning to Humanity," reprinted in Henry W. Kendall, *A Distant Light: Scientists and Public Policy* (New York: Springer, 2000), 198–201.
8. Union of Concerned Scientists, "World Scientists' Warning to Humanity," 198.

9. Union of Concerned Scientists, "World Scientists' Warning to Humanity," 199.
10. Union of Concerned Scientists, "World Scientists' Warning to Humanity," 201.
11. Union of Concerned Scientists, "World Scientists' Warning to Humanity," 201.
12. Dipesh Chakrabarty, "The Climate of History: Four Theses," *Critical Inquiry* 35, no. 2 (2009): 221.
13. Andreas Malm and Alf Hornborg, "The Geology of Mankind? A Critique of the Anthropocene Narrative," *The Anthropocene Review* 1, no. 1 (2014): 66–67. It seems to me that the conflict here is not irreconcilable. I think that Malm and Hornborg may be right when claiming that there will be lifeboats for the rich and privileged "as long as there are human societies on Earth." But the prospect of a runaway Anthropocene—the prospect that rewrites disciplinary concerns and codes—is precisely that there may be no human societies on Earth in the future. And this does not necessarily compel one to think of the ultimate prospect of human extinction due to altering planetary conditions to the extent that they no longer support human life. To see why Chakrabarty may eventually be right even in light of the criticism, it is enough to consider a potential societal collapse. For as soon as there are no functioning societies, there are no socially privileged individuals who could have anything like a lifeboat.
14. For a sample of recent critiques and discussions of anthropocentrism, human exceptionalism, and human expansionism across disciplines in the humanities, see *Anthropocentrism: Humans, Animals, Environments*, ed. Rob Boddice (Leiden: Brill, 2011); *The Nonhuman Turn*, ed. Richard Grusin (Minneapolis: University of Minnesota Press: 2015); Veronica Strang, "The Gaia Complex: Ethical Challenges to an Anthropocentric 'Common Future,'" in *The Anthropology of Sustainability: Beyond Development and Progress*, eds. Marc Brightman and Jerome Lewis (New York: Palgrave, 2017), 207–228; Jeremy A. Ross, "Durkheim and the Homo Duplex: Anthropocentrism in Sociology," *Sociological Spectrum* 37, no. 1 (2017): 18–26; Joyce E. Chaplin, "Can the Nonhuman Speak? Breaking the Chain of Being in the Anthropocene," *Journal of the History of Ideas* 48, no. 4 (2017): 509–529; Eileen Crist and Helen Kopnina, "Unsettling Anthropocentrism," *Dialectical Anthropology* 38, no. 4 (2014): 387–396. Critical posthumanism advocates an anti-, post-, or non-anthropocentric stance since about two decades. I will touch upon this at a later point of this chapter. In environmental ethics, there has also been a larger debate on anthropocentrism, peaking in the 1980s, which recent discussions typically do not address.

15. Eileen Crist, "Reimagining the Human," *Science* 362, Issue 6420 (2018): 1242. In her recent book, Crist makes clear that she prefers to use the term "human supremacy" over "anthropocentrism" because she thinks of the latter as an apolitical concept, even if the two terms can be used as synonyms. Eileen Crist, *Abundant Earth: Toward and Ecological Civilization* (Chicago: University of Chicago Press, 2019), 44.
16. Crist, "Reimagining the Human," 1243.
17. Crist, "Reimagining the Human," 1243.
18. Union of Concerned Scientists, "World Scientists' Warning to Humanity," 201.
19. Union of Concerned Scientists, "World Scientists' Warning to Humanity," 200.
20. Crist, "Reimagining the Human," 1243.
21. Crist, "Reimagining the Human," 1243.
22. Union of Concerned Scientists, "World Scientists' Warning to Humanity," 200.
23. Union of Concerned Scientists, "World Scientists' Warning to Humanity," 201.
24. Crist, "Reimagining the Human," 1244.
25. Kolbert, *The Sixth Extinction*, 268.
26. To be clear, none of this amounts to a defense of anthropocentrism. The only thing I claim is that there is an extent to which even anti-anthropocentric agendas necessarily remain anthropocentric.
27. To name a few approaches, see Nick Bostrom, *Superintelligence: Paths, Dangers, Strategies* (Oxford: Oxford University Press, 2014); Nicholas Agar, *Truly Human Enhancement: A Philosophical Defense of Limits* (Cambridge, MA: MIT Press, 2014); Sheila Jasanoff, *The Ethics of Invention: Technology and the Human Future* (New York: W. W. Norton and Company, 2016).
28. Martin Rees, *Our Final Hour: A Scientist's Warning: How Terror, Error, and Environmental Disaster Threaten Humankind's Future in this Century – On Earth and Beyond* (New York: Basic Books, 2003), vii.
29. For the latest Doomsday Clock Statement see John Mecklin (ed.), "It Is 100 Seconds to Midnight: 2020 Doomsday Clock Statement," Science and Security Board, *Bulletin of the Atomic Scientists* (2020). Available at: https://thebulletin.org/wp-content/uploads/2020/01/2020-Doomsday-Clock-statement.pdf
30. Jenny Andersson, *The Future of the World: Futurology, Futurists, and the Struggle for the Post-Cold War Imagination* (Oxford: Oxford University Press, 2018).
31. Andersson, *The Future of the World*, 2.
32. Andersson, *The Future of the World*, 48.

33. John Asafu-Adjaye et al., *An Ecomodernist Manifesto* (2015), 17. Available at: https://static1.squarespace.com/static/5515d9f9e4b04d5c3198b7bb/t/552d37bbe4b07a7dd69fcdbb/1429026747046/An+Ecomodernist+Manifesto.pdf
34. For an overview of transhumanist views see *The Transhumanist Reader*, eds. Max More and Natasha Vita-More (Malden: Wiley-Blackwell, 2013). For situating transhumanism with other contemporary intellectual movements that aim at a redefinition of the human see Francesca Ferrando, "Posthumanism, Transhumanism, Antihumanism, Metahumanism, and New Materialisms: Differences and Relations," *Existenz* 8, no. 2 (2013): 26–32; and Tamar Sharon, *Human Nature in an Age of Biotechnology: The Case for Mediated Posthumanism* (Dordrecht: Springer, 2014).
35. For a discussion of the specificity of post-Second World War technology see Chapter 2.
36. The humanities and the social sciences have already developed various ways to interpret and criticize today's catastrophe-dominated thinking and extinction-talk, with respect to both humans and nonhumans. See, for instance, Claire Colebrook, *Death of the PostHuman: Essays on Extinction, Vol 1* (Open Humanities Press, 2014); Nigel Clark, "Geo-politics and the Disaster of the Anthropocene," *The Sociological Review* 62, no. S1 (2014), 19–37; Stefan Skrimshire, "Climate Change and Apocalyptic Faith," *Wiley Interdisciplinary Reviews: Climate Change* 5, no. 2 (2014): 233–246; Isabelle Stengers, *In Catastrophic Times: Resisting the Coming Barbarism* (Open Humanities Press, 2015); Ursula K. Heise, *Imagining Extinction: The Cultural Meanings of Endangered Species* (Chicago: University of Chicago Press, 2016); Eva Horn, *The Future as Catastrophe: Imagining Disaster in the Modern Age*, trans. Valentine Pakis (New York: Columbia University Press, 2018).
37. For a debate on the prospect of a technological singularity and on the feasibility of the notion see *Singularity Hypotheses: A Scientific and Philosophical Assessment*, eds. Amnon H. Eden, James H. Moor, Johnny H. Søraker and Eric Steinhart (Berlin and Heidelberg: Springer, 2012).
38. Vernor Vinge, "The Coming Technological Singularity: How to Survive in the Post-Human Era," in *Vision-21: Interdisciplinary Science and Engineering in the Era of Cyberspace*, Proceedings of a symposium cosponsored by the NASA Lewis Research Center and the Ohio Aerospace Institute, Westlake, Ohio, March 30–31 (1993), 12–13.
39. Johan Rockström et al., "Planetary Boundaries: Exploring the Safe Operating Space for Humanity," *Ecology and Society* 14, no. 2 (2009): art. 32; Will Steffen et al., "Planetary Boundaries: Guiding Human Development on a Changing Planet," *Science* 347, no. 6223 (2015): 1259855.

40. Steffen et al., "Planetary Boundaries," 1.
41. For the most influential version of anti- or postanthropocentric critical posthumanism see Rosi Braidotti, *The Posthuman* (Cambridge, Polity, 2013). See also Francesca Ferrando, *Philosophical Posthumanism* (London: Bloomsbury, 2019). For the possibility of an anti-anthropocentric posthumanist history see Ewa Domanska, "Posthumanist History," in *Debating New Approaches to History*, eds. Marek Tamm and Peter Burke (London: Bloomsbury, 2018), 327–338. See also the discussion of anthropocentrism and the references in the earlier pages of this chapter.
42. See Zoltán Boldizsár Simon, "Two Cultures of the Posthuman Future," *History and Theory* 58, no. 2 (2019): 171–184.
43. Fukuyama, *Our Posthuman Future*.
44. See Baridotti, *The Posthuman*, 143–185. More recently see Rosi Braidotti, "A Theoretical Framework for the Critical Posthumanities," *Theory, Culture & Society* 36, no. 6 (2019): 31–61.
45. Georges-Louis LeClerc, Le Comte de Buffon, *The Epochs of Nature*, trans. and eds. Jan Zalasiewicz, Anne-Sophie Milon, and Mateusz Zalasiewicz (Chicago: University of Chicago Press, 2018).
46. Jan Zalasiewicz, Sverker Sörlin, Libby Robin, and Jacques Grinewald, "Introduction: Buffon and the History of the Earth," in Georges-Louis LeClerc, Le Comte de Buffon, *The Epochs of Nature*, trans. and ed. Jan Zalasiewicz, Anne-Sophie Milon, and Mateusz Zalasiewicz, Chicago: University of Chicago Press, 2018, xiv.
47. For overviews of different positions, see Bruce D. Smith and Melinda A. Zeder, "The Onset of the Anthropocene," *Anthropocene* 4 (2013): 8–13; Simon L. Lewis and Mark A. Maslin, "Defining the Anthropocene," *Nature* 519 Issue 7542 (2015): 171–180.
48. Jan Zalasiewicz et al., "The Working Group on the Anthropocene: Summary of Evidence and Interim Recommendations," *Anthropocene* 19 (2017): 58.
49. With or without direct links to the Anthropocene debate, see, for instance, Lynn Hunt, *Measuring Time, Making History* (Budapest: Central European University Press, 2008); *Breaking Up Time: Negotiating the Borders between Present, Past and Future*, eds. Chris Lorenz and Berber Bevernage (Göttingen: Vanderhoeck & Ruprecht, 2013); Stefan Tanaka, "History without Chronology," *Public Culture* 28, no. 1 (2015), 161–186; Victoria Fareld, "(In) Between the Living and the Dead: New Perspectives on Time in History," *History Compass* 14, no. 9 (2016): 430–440; Chris Lorenz, "The Times They Are a-Changin': On Time, Space and Periodization in History," in *Palgrave Handbook of Research in Historical Culture and Education*, eds. Mario Carretero, Stefan Berger and Maria Grever (London: Palgrave, 2017), 109–131; *Rethinking Historical Time: New Approaches to*

Presentism, eds. Marek Tamm and Laurent Olivier (London: Bloomsbury, 2019): see especially the chapter Helge Jordheim, "Return to Chronology," in *Rethinking Historical Time: New Approaches to Presentism*, eds. Marek Tamm and Laurent Olivier (London: Bloomsbury, 2019), 43–56. Finally, for general discussions of periodization by eminent historians see Ludmilla Jordanova, *History in Practice*, 2nd ed. (London: Bloomsbury, 2007), 105–125; and Jacques Le Goff, *Must We Divide History Into Periods?* trans. M. B. DeBevoise (New York: Columbia University Press, 2014).

50. Dipesh Chakrabarty, *The Human Condition in the Anthropocene*, The Tanner Lectures in Human Values, delivered at Yale University (February 18–192,015), 137–188. Available at: https://tannerlectures.utah.edu/Chakrabarty%20manuscript.pdf
51. Karl Jaspers, *Man in the Modern Age*, trans. Eden and Cedar Paul (New York: Henry Holt and Company, 1933); Karl Jaspers, *The Atom Bomb and the Future of Man*, trans. E. B. Ashton (Chicago: University of Chicago Press, 1963).
52. Chakrabarty, *The Human Condition in the Anthropocene*, 144.
53. See, especially, Chapter 3 in Zoltán Boldizsár Simon, *History in Times of Unprecedented Change: A Theory for the 21st Century* (London: Bloomsbury, 2019); Zoltán Boldizsár Simon, "The Transformation of Historical Time: Processual and Evental Temporalities," in *Rethinking Historical Time: New Approaches to Presentism*, eds. Marek Tamm and Laurent Olivier (London: Bloomsbury, 2019), 71–84.

Bibliography

Agar, Nicholas. *Truly Human Enhancement: A Philosophical Defense of Limits*. Cambridge, MA: MIT Press, 2014.

Andersson, Jenny. *The Future of the World: Futurology, Futurists, and the Struggle for the Post-Cold War Imagination*. Oxford: Oxford University Press, 2018.

Asafu-Adjaye, John et al. *An Ecomodernist Manifesto*. 2015. Available at: https://static1.squarespace.com/static/5515d9f9e4b04d5c3198b7bb/t/552d37bbe4b07a7dd69fcdbb/1429026747046/An+Ecomodernist+Manifesto.pdf

Boddice, Rob, ed. *Anthropocentrism: Humans, Animals, Environments*. Leiden: Brill, 2011.

Bostrom, Nick. *Superintelligence: Paths, Dangers, Strategies*. Oxford: Oxford University Press, 2014.

Braidotti, Rosi. *The Posthuman*. Cambridge: Polity, 2013.

Braidotti, Rosi 2018. "A Theoretical Framework for the Critical Posthumanities." *Theory, Culture & Society* 36, no. 6 (2019): 31–61.

Buffon, Le Comte de, Georges-Louis LeClerc. *The Epochs of Nature*. Translated and edited by Jan Zalasiewicz, Anne-Sophie Milon, and Mateusz Zalasiewicz, Chicago: University of Chicago Press, 2018.

Chakrabarty, Dipesh. "The Climate of History: Four Theses." *Critical Inquiry* 35, no 2 (2009): 197–222.

Chakrabarty, Dipesh. *The Human Condition in the Anthropocene*. The Tanner Lectures in Human Values, delivered at Yale University, 18–19 February, 2015. 137–188. Available at: https://tannerlectures.utah.edu/Chakrabarty%20manuscript.pdf

Chaplin, Joyce E. "Can the Nonhuman Speak? Breaking the Chain of Being in the Anthropocene." *Journal of the History of Ideas* 48, no 4 (2017): 509–529.

Clark, Nigel. "Geo-politics and the Disaster of the Anthropocene." *The Sociological Review* 62, no. S1 (2014): 19–37.

Colebrook, Claire. "Framing the End of Species." *symplokē* 21, no. 1–2 (2013): 51–63.

Colebrook, Claire. *Death of the PostHuman: Essays on Extinction, Vol 1*. Open Humanities Press, 2014.

Crist, Eileen, and Helen Kopnina, "Unsettling Anthropocentrism." *Dialectical Anthropology* 38, no. 4 (2014): 387–396.

Crist, Eileen. "Reimagining the Human." *Science* 362, Issue 6420 (2018): 1242–1244.

Crist, Eileen. *Abundant Earth: Toward and Ecological Civilization*, Chicago: University of Chicago Press, 2019.

De Vos, Jurriaan M. et al. "Estimating the Normal Background Rate of Species Extinction," *Conservation Biology* 29, no. 2 (2014): 452–462.

Domanska, Ewa. "Posthumanist History." In *Debating New Approaches to History*, edited by Marek Tamm and Peter Burke, 327–338. London: Bloomsbury, 2018.

Eden, Amnon H., James H. Moor, Johnny H. Søraker and Eric Steinhart, eds. *Singularity Hypotheses: A Scientific and Philosophical Assessment*. Berlin and Heidelberg: Springer, 2012.

Fareld, Victoria. "(In) Between the Living and the Dead: New Perspectives on Time in History." *History Compass* 14, no. 9 (2016): 430–440.

Ferrando, Francesca. "Posthumanism, Transhumanism, Antihumanism, Metahumanism, and New Materialisms: Differences and Relations." *Existenz* 8, no. 2 (2013): 26–32.

Ferrando, Francesca. *Philosophical Posthumanism*. London: Bloomsbury, 2019.

Fukuyama, Francis. *Our Posthuman Future: Consequences of the Biotechnology Revolution*. New York: Farrar, Straus and Giroux, 2002.

Grusin, Richard, ed. *The Nonhuman Turn*. Minneapolis: University of Minnesota Press, 2015.

Heise, Ursula K. *Imagining Extinction: The Cultural Meanings of Endangered Species*, Chicago: University of Chicago Press, 2016.

Horn, Eva. *The Future as Catastrophe: Imagining Disaster in the Modern Age*. Translated by Valentine Pakis. New York: Columbia University Press, 2018.
Hunt, Lynn. *Measuring Time, Making History*. Budapest: Central European University Press, 2008.
Jasanoff, Sheila. *The Ethics of Invention: Technology and the Human Future*. New York: W. W. Norton and Company, 2016.
Jaspers, Karl. *Man in the Modern Age*. Translated by Eden and Cedar Paul. New York: Henry Holt and Company, 1933.
Jaspers, Karl. *The Atom Bomb and the Future of Man*. Translated by E. B. Ashton. Chicago: University of Chicago Press, 1963.
Jordanova, Ludmilla. *History in Practice*. 2nd ed. London: Bloomsbury, 2006.
Jordheim, Helge. "Return to Chronology." In *Rethinking Historical Time: New Approaches to Presentism*, edited by Marek Tamm and Laurent Olivier, 43–56. London: Bloomsbury, 2019.
Kolbert, Elizabeth. *The Sixth Extinction: An Unnatural History*. London: Bloomsbury, 2014.
Le Goff, Jacques. *Must We Divide History Into Periods?* Translated by M. B. DeBevoise. New York: Columbia University Press, 2014.
Leakey, Richard, and Roger Lewin. *The Sixth Extinction: Patterns of Life and the Future of Humankind*. New York: Anchor Books, 1996.
Lewis, Simon L., and Mark A. Maslin. "Defining the Anthropocene." *Nature* 519, Issue 7542 (2015): 171–180.
Lorenz, Chris. "The Times They Are a-Changin': On Time, Space and Periodization in History." In *Palgrave Handbook of Research in Historical Culture and Education*, edited by Mario Carretero, Stefan Berger and Maria Grever, 109–131. London: Palgrave, 2017.
Lorenz, Chris, and Berber Bevernage, eds. *Breaking Up Time: Negotiating the Borders between Present, Past and Future*. Göttingen: Vanderhoeck & Ruprecht, 2013.
Malm, Andreas, and Alf Hornborg. "The Geology of Mankind? A Critique of the Anthropocene Narrative." *The Anthropocene Review* 1, no. 1 (2014): 62–69.
Mecklin, John, ed. "It Is 100 Seconds to Midnight: 2020 Doomsday Clock Statement." Science and Security Board, *Bulletin of the Atomic Scientists* (2020). Available at: https://thebulletin.org/wp-content/uploads/2020/01/2020-Doomsday-Clock-statement.pdf
More, Max, and Natasha Vita-More, eds. *The Transhumanist Reader*. Malden: Wiley-Blackwell, 2013.
Pimm, Stuart J., and Thomas M. Brooks. "The Sixth Extinction: How Large, Where, and When?," In *Nature and Human Society: The Quest for a Sustainable World*, edited by Peter H. Raven and Tania Williams, 46–62. Washington, D.C.: National Academy Press, 2000.

Rees, Martin. *Our Final Hour: A Scientist's Warning: How Terror, Error, and Environmental Disaster Threaten Humankind's Future in this Century – On Earth and Beyond*. New York: Basic Books, 2013.
Ripple, William J. et al. "World Scientists' Warning to Humanity: A Second Notice." *BioScience* 67, no. 12 (2017): 1026–1028.
Rockström, Johan et al. "Planetary Boundaries: Exploring the Safe Operating Space for Humanity." *Ecology and Society* 14, no. 2 (2009): art. 32.
Ross, Jeremy A. "Durkheim and the Homo Duplex: Anthropocentrism in Sociology." *Sociological Spectrum* 37, no. 1 (2017): 18–26.
Sandel, Michael J. *The Case Against Perfection: Ethics in the Age of Genetic Engineering*. Cambridge, MA: The Belknap Press of Harvard University Press, 2007.
Sharon, Tamar. *Human Nature in an Age of Biotechnology: The Case for Mediated Posthumanism*. Dordrecht: Springer, 2014.
Simon, Zoltán Boldizsár. "Two Cultures of the Posthuman Future." *History and Theory* 58, no. 2 (2019): 171–184.
Simon, Zoltán Boldizsár. "The Transformation of Historical Time: Processual and Evental Temporalities." In *Rethinking Historical Time: New Approaches to Presentism*, edited by Marek Tamm and Laurent Olivier, 71–84. London: Bloomsbury, 2019.
Simon, Zoltán Boldizsár. *History in Times of Unprecedented Change: A Theory for the 21st Century*. London: Bloomsbury, 2019.
Skrimshire, Stefan. "Climate Change and Apocalyptic Faith." *Wiley Interdisciplinary Reviews: Climate Change* 5, no. 2 (2014): 233–246.
Smith, Bruce D., and Melinda A. Zeder. "The Onset of the Anthropocene." *Anthropocene* 4 (2013): 8–13.
Steffen, Will et al. "Planetary Boundaries: Guiding Human Development on a Changing Planet." *Science* 347, Issue 6223 (2015): 1259855.
Stengers, Isabelle. *In Catastrophic Times: Resisting the Coming Barbarism*. Open Humanities Press, 2015.
Strang, Veronica. "The Gaia Complex: Ethical Challenges to an Anthropocentric 'Common Future.'" In *The Anthropology of Sustainability: Beyond Development and Progress*, edited by Marc Brightman and Jerome Lewis, 207–228. New York: Palgrave, 2017.
Tamm, Marek, and Laurent Olivier, eds. *Rethinking Historical Time: New Approaches to Presentism*. London: Bloomsbury, 2019.
Tanaka, Stefan. "History without Chronology." *Public Culture* 28, no. 1 (2015): 161–186.
Toivanen T. et al. "The Many Anthropocenes: A Transdisciplinary Challenge for the Anthropocene Research." *The Anthropocene Review* 4, no 3 (2017): 183–198.
Union of Concerned Scientists. "World Scientists' Warning to Humanity." [1992] In Henry W. Kendall, *A Distant Light: Scientists and Public Policy*, 198–201. New York: Springer, 2000.

Vinge, Vernor. "The Coming Technological Singularity: How to Survive in the Post-Human Era." In *Vision-21: Interdisciplinary Science and Engineering in the Era of Cyberspace*, 11–22. Proceedings of a Symposium Cosponsored by the NASA Lewis Research Center and the Ohio Aerospace Institute, Westlake, Ohio, 30–31 March 1993.

Zalasiewicz, Jan et al. "The Working Group on the Anthropocene: Summary of Evidence and Interim Recommendations." *Anthropocene* 19 (2017): 55–60.

Zalasiewicz, Jan, Sverker Sörlin, Libby Robin, and Jacques Grinewald, "Introduction: Buffon and the History of the Earth." In Georges-Louis LeClerc, Le Comte de Buffon, *The Epochs of Nature*, translated and edited by Jan Zalasiewicz, Anne-Sophie Milon, and Mateusz Zalasiewicz, xiii–xxxiv. Chicago: University of Chicago Press, 2018.

CHAPTER 5

The Historical Event

Abstract The chapter reviews notions of the historical event as representatives of the closest available conceptual category to which the new instances of the epochal—the sixth mass extinction event, the technological singularity, and the potential transgression of planetary boundaries—can be measured. Special attention is given to three notions, developed, respectively, by Rolf Gruner (in analytic philosophy), Hayden White (in historical theory), and William Sewell (in history, political science, historical sociology). The discussion of the chapter is focused on two heavily interrelated aspects instrumental in situating the notion of the historical event with the epochal in the next chapter: first, the relationship between event and novelty, and second, the transformative potential of historical events.

Keywords Historical event • Novelty • Change • Transformative potential

Notions of the Historical Event

This book opened with the claim that the concept of the historical event is a category far too narrow to adequately capture the character of certain events of late. The subsequent chapters sketched the emergence of a peculiar kind of epochal thought and introduced three prominent instances of

Z. B. Simon, *The Epochal Event*, Palgrave Studies in the History of Science and Technology,
https://doi.org/10.1007/978-3-030-47805-6_5

the new epochal: the sixth mass extinction event, the technological singularity, and the potential transgression of planetary boundaries. The discussion of a new epochality already hinted at the two most crucial aspects in which the category of a specifically "historical" event may indeed be narrow. First, in light of the human-technology-nature entanglement, any notion of the historical event with a scope limited to the world of human affairs seems far too constricted. Second, when the most momentous transformative events are either perceived as ongoing ones or appear as potential future occurrences, notions of the historical event that typically—if not exclusively—refer to events that already happened come out, again, as hopelessly limited.

At this point, however, the narrowness of the historical event is only a claim yet to be supported, and the notion of the epochal event has not gained any conceptual clarity so far. In addressing these questions, this chapter intends to discuss various notions of the historical event, only in order to enable the next chapter to pull together the diverging threads of this book and conceptualize the epochal event by situating the new instances of the epochal with existing notions of the historical event.

The question of historical events has been addressed in various ways in post-Second World War historiography and historical theory. Theoretically inclined historians and philosophers of history equally felt compelled to comment on the role historical events play in the work of professional historical studies in particular and in historical thinking in general. Rather obviously, not all approaches to the historical event understand the notion in a way that could be relevant for this book: not all of them associate the concept with historical change, and not all of them consider events as momentous occurrences that signal large-scale transformations.

To begin with, *talking about* the role of historical events in historical thinking does not necessarily mean *having a concept* or a notion of the historical event. Historiographical debates may be the best examples in this respect. Postwar French historiography was especially keen on talking a lot about historical events, but typically without the intent of developing a category of historical thought. Historiographical debates in France about historical events simply had an entirely different subject matter and an imperative other than theorizing a notion of transformative events. They were typically occupied with the question of the feasibility of writing a "history of events."

The terms of the discussion were set by Fernand Braudel in the middle of the last century, as part of a larger program that attempted to move away from a history of events to what Braudel called the history of the *longue durée*. In making the case for the latter, Braudel associated events with the "short time span," and his understanding of the event meant something "explosive" with a "delusive smoke" that "fills the minds of its contemporaries, but it does not last, and its flame can scarcely ever be discerned."[1] The "history of events" context dominated subsequent twentieth-century historiographical discussions, challenging or endorsing the Braudelian view on the role and significance of historical events. It informed the contrast to social histories interested in exploring a larger-scale history of structures, and remained the general framework of historiographical exchange even when an event-oriented approach in historical writing seemed to return in the last decades of the previous century as linked with the question of narrative in historiography.[2]

The "revival of narrative" in historical studies coincided with the heyday of the work of theories and philosophies of history on the question of history as narrative. Whether there is a direct link between the two discourses or they rather talked past each other is a question that I do not wish to explore here. More important is to say a few words about how the question of historical events featured in historical theory in the heyday of narrativism. Generally speaking, for the theory and philosophy of history of the second half of the last century, historical events mattered inasmuch as they formed a sequence within a narrative structure associated with historical writing. Despite their respective focuses and theoretical disagreements, analytic philosophies, phenomenological approaches, and Hayden White's literary theory-inspired narrative philosophy of history have all been on the same platform at least about this kind of (un)interest in historical events.[3]

There nevertheless are notable exceptions that paid genuine attention to the notion as a category of historical thought. As to analytic philosophy of history, it was chiefly invested in exploring how narratives explain historical events without the practical need to elevate the historical event into a concept. Yet there is an exception here, an article from Rolf Gruner, whose writings did not have an impact measurable to the ideas of fellow analytic philosophers of history, such as Arthur Danto, William Walsh, Walter Gallie, or Louis Mink. In 1969, it was nevertheless Gruner who devoted an article to the notion of the historical event.[4] As to Hayden White, it is not his narrative philosophy of history but two of his later essays that engage with the theme in a profound manner. In the first one from the end of the 1990s, White reflects on what he calls the "modernist event." The second, written more than a decade later, is explicitly devoted to the notion of the historical event. Terminological differences aside, White actually talks about the same thing in both essays: the "modernist event" as a twentieth-century historical event.[5]

This chapter will take a closer look at the ideas of both Gruner and White, along with a third and in fact most useful conceptualization of the historical event. This would be William Sewell's notion of the historical event as the harbinger of structural changes.[6] In taking the promised closer look, this chapter will focus on two heavily interrelated aspects that may be the most instrumental in situating the historical event with instances of the epochal (explored in the previous chapter): first, the relationship between event and novelty, and second, the transformative potential of historical events.

Historical Event and Novelty

Gruner's analysis of the historical event aims at bringing to light the basic constitution of the notion as it appears in common usage. Given the fact that Gruner's article begins with the sentence claiming that "no historian spends sleepless nights over the question 'What is an historical event?',"[7] it is safe to assume that a general lack of conceptual reflection on the notion within historical studies (which does not mean of course a complete lack) leads Gruner to believe that there is little difference between the typical usage of the notion in historical writing and in everyday speech. Regardless of the question of whether history books and everyday speech indeed use the notion of the historical event in a similar fashion, Gruner states that the fundamental tenet of any event of a historical character is that it ushers in a change: "a certain state is prevailing; an event occurs; and afterwards the state is different."[8]

With this point of departure, Gruner touches upon several derivative points in his analysis. I would like to highlight two of them. The first is that the change inaugurated by the historical event needs a constant, and Gruner identifies this constant with the self-identity of the object that is described as going through changes. According to Gruner, "if an object changes its identity it is no longer the same object, and if it is no longer the same object then we involve ourselves in a self-contradiction, i.e., the word 'change' becomes inapplicable."[9] This is a highly questionable contention. Let's keep it in mind and save it for the upcoming discussion of the epochal event.

The second point to highlight concerns an inherent relativism in the notion of the historical event. Gruner claims that it depends on the context—that is, it depends on us—if we see something as a historical event in the first place. In the words of Gruner, "what is viewed in a given case as an event depends on what is viewed as the state which is, or might be, changed by it." Besides, "it is likewise a matter of choice what exactly is viewed as being in that state, i.e., what constitutes the object in question."[10] Gruner illuminates the relativism of the historical event through discussing the battle of Waterloo. Historians are not only "free to describe" the battle as "affecting the state of France, or of Europe, or of England, or of the world"; they can also consider it something other than an event. They might as well regard the battle to be the object itself that they investigate. Historians might just as well attempt to "describe how the battle developed from beginning to end," recount "the phases through which it passed," and "account for the change from one phase to another by reference to subevents of the battle."[11]

As it stands, Gruner is not alone in thinking that relativism is inescapable when it comes to identifying an occurrence as a historical event. Sewell makes the same point in considering the taking of the Bastille during the French Revolution as an example of a historical event worthy of its name, that is, an event that brings about structural changes. Like Gruner, Sewell thinks that conceiving of something as an event requires a "judgment" about what is seen as an event in the first place. Given that the scale and scope of what we see as historical events may largely vary, and given that we may see smaller events within bigger events, Sewell puts forward the claim that events have a "fractal character": historical events are composed of "sub-events," events within larger events. The taking of the Bastille on July 14 in 1789 is, according to Sewell, just as much an event as is "the king's entry to into Paris on July 17," or "his actions on the balcony of the Hotel de Ville that afternoon," or "the French Revolution as a whole."[12]

Unlike Gruner, however, Sewell does not look at the changes inaugurated by historical events in developmental terms. He thinks that "when changes do take place, they are rarely smooth and linear in character"; instead, they "tend to be clustered into relatively intense bursts."[13] Associating events with intense bursts that defy the logic of developmental processes means that Sewell advocates an account of historical events inseparable from the advent of radical novelty. The extremity of the novelty springing out of historical events is provided by the unpredictability of the induced transformations as based on the knowledge of the way that led to the events themselves. In Sewell's words, historical events "tend to transform social relations in ways that could not be fully predicted from the gradual changes that made them possible."[14] With respect to the taking of the Bastille, this means that Sewell looks at the event "from a particular perspective" that emphasizes the novelty that the event brings about. Differently put, he looks at historical events from the viewpoint of their consequences to decide whether an occurrence qualifies as a historical event in the first place. And if the taking of the Bastille does qualify, it is because it ushered in a novelty: the articulation of "popular violence with the nation's sovereign will in the new concept of revolution."[15]

Defining historical events in consequentialist terms has implications concerning their duration: they must end when the genuine novelty in which they result comes to existence. Accordingly, Sewell assigns a duration to the historical event of the taking of the Bastille—between July 14 and July 23—through identifying its end date as the point at which the National Assembly already referred to the event in terms of the novel concept of revolution. All in all, the taking of the Bastille qualifies as a historical event inasmuch as it brings about the political notion of revolution, which, prior to the event, simply did not exist as one that refers to legitimate but violent transformations of political systems. It did not exist before as distinct from "illegitimate" forms of popular violence which occurred in the days following the taking of the Bastille, from which the National Assembly distinguished the "legitimate" form though labeling it "revolution."

Sewell elaborates on several aspects of historical events that I do not wish to recount here. His explorations range from the pre-conditions of events to their ritualistic character. The aspect I want to take forward to the upcoming discussion of the epochal event is the one which contradicts Gruner's earlier point concerning the necessity of a constant provided by an object that changes only in its states but not in its identity. Unlike Gruner, Sewell associates the change resulting from the event with the coming about of something previously nonexistent. He states this point most clearly in a discussion of historical events as "cultural transformations":

> The novel articulation that makes this happening a momentous event in world history is an act of signification. Terms – for example, "Bastille" and "revolution," but also "people," "liberty," "despotism," and so on – took

> on authoritative new meanings that, taken together, reshaped the political world. This implies that events are, literally, significant: they signify something new and surprising. They introduce new conceptions of what really exists (the violent crowd as the people's will in action), of what is good (the people in ecstatic union), and of what is possible (revolution, a new kind of regeneration of the state and the nation). The most profound consequence of the taking of the Bastille was, then, a reconstruction of the very categories of French political culture and political action.[16]

The takeaway message of the quoted passage is that historical events bring about "new conceptions of what really exists." In Sewell's version, the historical event has the potential to recast the ways in which we conceive of the world and ourselves by creating the concepts and categories through which we make sense of the human world.

Let's call this the transformative potential of historical events, and let's define this potential as the perceived capacity of historical events to bring about radical novelty in human affairs. In the remaining pages of this chapter, I would like to dwell a bit more on the question of this potential through a criticism of Hayden White's notion of the historical event.

The Transformative Potential of Historical Events

In his two essays on the modernist event as a twentieth-century version of the historical event, White asks a question that may imply even more spectacular changes than those outlined by Sewell. It goes as follows: "Can we imagine a new *kind* of event breaking in on our own world which might manifest evidence of another, alternative system of existence that differs utterly from our own?"[17]

White's question seems to imply a task that suspiciously resembles the task of this book, that is, to conceptualize a new kind of event. Can it be that White was already pondering issues directly relevant for what I call here as the epochal event? What's more, can it be that White's modernist event and my epochal event are only different names for the same phenomenon? The suspicion that this may be the case could find support in the paragraph preceding White's question. In that paragraph, White muses over the way in which "weapons of mass destruction cause a quantum leap in the *history* of warfare," while "antibiotics and genetic engineering change definitely the nature of health care for the foreseeable future."[18] And there is more. At some point, White even remarks that "certain events in modernity – space travel, genetic engineering, atomic weaponry – are so utterly different from anything previously thought possible that even a *modern* peasant or bourgeois might be forgiven for taking them as 'miracle.'"[19] So what to make of this?

Despite the fact that there most certainly is an overlap between the phenomena upon which both White and this book reflect, it would be completely misleading to interpret White as having an idea of how some of the examples he lists may also be instances of a specifically postwar perception of large-scale anthropogenic changes in an entangled human-technological-natural world that result in epochal transformations. The kind of event White is interested in cannot have much to do with the new kind of epochal thought, because it does not involve the entanglement of the human, the technological, and the natural worlds.

Instead, White considers his examples as having a "traumatic" character, and he ends up subsuming such traumatic events under the category of the "modernist event." Most importantly, all this provides yet another occasion for White to struggle with the very same sets of questions to which he devoted an exceptional scholarly career. In White's view, the main question posed by modernist traumatic events is that of how to represent them adequately in historical writing. In attempting to answer the question, it seems to me that White conflates space travel and genetic engineering with the Holocaust and treats them all as instances of the same generic phenomena. As I will show later, this treatment amounts to a category mistake. Yet, White's theory of the historical event as traumatic modernist event, precisely because of the category mistake it entails, may be instructive in indicating what an epochal event is most certainly not.

Although White must be credited with sensing that something is going on with the way we conceive of certain events of late, it was simply impossible for him to grasp the situation within the framework to which he remained committed since the publication of his groundbreaking essay "The Burden of History" in 1966.[20] It is now more than half a century since White began to argue that the discipline of history is stuck in ways of literary meaning-making inherited from nineteenth-century literary realism. Three decades after "The Burden of History," in his essay on the modernist event in 1996, White picked up his earlier thread of thought by claiming that the advent of a new kind of modernist events poses a challenge to such literary realism as a representational form.

Now, why is this the case? What is so peculiar about "modernist" events? To begin with, "modernist" and "traumatic" are not the only adjectives White uses to describe the events of the twentieth century. He also calls such events "holocaustal," referring to events "which not only could not possibly have occurred before the twentieth century but the nature, scope, and implications of which no prior age could ever have imagined."[21] The following passage indicates the diversity of the events White attempts to subsume under the category:

> Some of these "holocaustal" events – such as the two World Wars, the Great Depression, a growth in world population hitherto unimaginable, poverty and hunger on a scale never before experienced, pollution of the ecosphere by nuclear explosions and the indiscriminate disposal of contaminants, programs of genocide undertaken by societies utilizing scientific technology and rationalized procedures of governance and warfare (of which the German genocide of 6,000,000 European Jews is paradigmatic) – function in the consciousness of certain social groups exactly as infantile traumas are conceived to function in the psyche of neurotic individuals. This means that they cannot be simply forgotten and put out of mind, but neither can they be adequately remembered; which is to say, clearly and unambiguously identified as to their meaning and contextualized in the group memory in such a way as to reduce the shadow they cast over the group's capacities to go into its present and envision a future free of their debilitating effects.[22]

White thinks that the representational strategies of historical writing as we know it are bankrupt when facing with events "modernist," "traumatic," and "holocaustal." "After modernism," he says, "when it comes to the task of storytelling, whether in historical or in literary writing, the traditional techniques of narration become unusable – except in parody."[23] As an example to support this claim, White points to the explosion of the space shuttle *Challenger* in 1986, broadcasted live. Notwithstanding how people witnessing the event craved for immediate answers, White thinks that "it appeared impossible to tell any single authoritative story about what actually happened – which meant that one could tell any number of possible stories about it."[24]

As much as this may be true, it is hard not to note immediately that the lack of authority to tell a single story and the resulting possibility of telling any number of stories have actually not much to do with the claim concerning the unusability of "traditional techniques of narration." What the relativity of stories that can be told about the *Challenger* indicates is not a collapse of story form but an abundance of stories that can be told. I will return to this and other questionable aspects of White's theory in a moment.

For now, let's follow through with White's approach, which leads to the question of whether there is any plausible way to deal with the representational difficulties posed by modernist events. White thinks there is, and again, his proposed solution to overcome difficulties is the same in 1996 as it was in "The Burden of History" essay in 1966: history needs to align with the representational modes of newer forms of literary writing. White suggests that "the kinds of anti-narrative non-stories produced by literary modernism offer the only prospect for adequate representations of the kind of unnatural events – including the Holocaust – that mark our era and distinguish it absolutely from all 'history' that has come before it."[25]

The proposed solution remains unchanged even when White returns to the question of the historical event in his last book, *The Practical Past*. Leaping into the twenty-first century, White restates his previous position, except that now it appears as an accomplished fact. In literary modernism, White states in 2014, "history, the historical event, and historicality itself are taken over by a new kind of writing which, for want of a better term, we may call postmodernist." What this means is that "a distinctively 'historical' way of accounting for the intervention of a 'new kind of writing' requires us to identify the new 'content' or phenomenon for the representation of which the new kind of writing is thought to be adequate." As White makes clear, he "already alluded to 'the modernist event' as such a content," and "the 'substance' of the 'content' of this new kind of event is provided in the historiotheticized idea of 'trauma.'"[26]

Now, what is the problem with the way White approaches the question of the historical event today? Well, there is more than one. In fact, there are far too many. First, there is a category mistake in which a rather diverse set of events seems to qualify as "modernist," "traumatic" and "holocaustal," without a hint of what shared quality could justify their coupling together. In the course of elaborating on the category, White alludes to so many events as being "modernist" that in the end almost any event whatsoever comes out as one. Consider the following example:

> In short, the threat posed by the representation of such events as the Holocaust, the Nazi Final Solution, by the assassination of a charismatic leader such as Kennedy or Martin Luther King or Gandhi, or by an event such as the destruction of the Challenger, which had been symbolically orchestrated to represent the aspirations of a whole community, is nothing

> other than the threat of turning these events into the subject-matter of a narrative. Telling a story, however truthful, about such traumatic events might very well provide a kind of "intellectual mastery" of the anxiety which memory of their occurrence may incite in an individual or a community. But precisely insofar as the story is identifiable *as a story*, it can provide no lasting "psychic mastery" of such events.[27]

In the quoted passage, even the assassination of statespersons joins the earlier mentioned instances of the modernist event, such as the Holocaust, the World Wars, the *Challenger* catastrophe, nuclear warfare, rapid global population growth, air pollution, space travel, genetic engineering, and so forth. The shared feature of these events cannot be the scale and scope that White refers to every now and then, because some of these are rather small events. Nor can it be their assumed difference from anything previously seen, because, again, some of them are rather common occurrences. Or is there a way in which twentieth-century assassinations significantly differ in character from the political assassination of the nineteenth century, which has been described by Rachel Hoffman as "the age of assassination"?[28] As it stands, White does not say anything about the specificity of the assassinations he lists.

The shared quality of the new kind of events cannot even be their traumatic "content" either, one of the main categories under which White attempts to subsume all. For how is space travel traumatic and to whom exactly? Even if we concede to White that technological prospects may be catastrophic in character (although not exclusively, as seen earlier), that which is catastrophic is still not necessarily traumatic. This is not to say that many of White's examples are not traumatic for those who lived through them and for a broader societal consciousness that remembers to them. The Holocaust and the World Wars most certainly are. The point I wish to make is that the entire set of diverse events White brings together cannot be united by a supposed traumatic content. Many of them are simply not experienced as personal psychological stress, let alone, as a societal equivalent of such psychological state.

The second problem with White's modernist event is connected to his interpretation of the *Challenger* catastrophe. It was supposed to support White's general claim concerning the collapse of narrative techniques in the representation of modernist events. Yet, what he ended up saying was only that witnessing the event enabled people to tell many narratives instead of a single authoritative one. Equating the collapse of storytelling techniques as adequate tools of sense-making with the fact that there may be many possible stories to tell about a historical event is, I think, yet another category mistake. Whereas the former claim is left unargued, the latter one is simply true of any event—even in light of White's own life oeuvre which intended to show that events can be narrativized in many ways and are in fact functions of historical narratives.[29]

Third, the on-the-spot shock of witnessing the *Challenger* catastrophe is the momentary collapse of cognitive capacities to attribute sense and meaning to whatever has just been witnessed, while historical writing is never on-the-spot. A temporary shock of first-hand experience and eye-witnessing in no way excludes to possibility of a post-evental narration. The question whether such post-evental narration *should* or *should not* be exercised on events experienced as defying pre-existing conceptualizations can of course be raised. But this is not White's question, and historians find manifold ways to narrativize several of the events White mentions as modernist ones.[30] True enough, this hardly happens from the point of view of a first person eye-witness, but again, historical writing is rarely about the personal experiences of historians. While those first-hand experiences are rather obviously effective in several ways in historiography—in terms of normative agendas, choosing subjects of study, raising questions, and so on—they are not what histories are typically *about*.

This is not to say that it would be utterly impossible to write about one's own first-hand experience in historiography. In a certain way, the rise of contemporary history offers a fairly good chance for historians to study what they themselves might have experienced, thereby providing a highly contested but increasingly popular option for historical scholarship.[31] The same applies to ego-histories—an innovation emerging in French historiography in the 1980s—that intend to recount the personal experiences of the historian in order to illuminate how they relate to the histories they author.[32] As options, however, even the most "contemporary" contemporary histories and the most recent ego-histories necessarily remain post-evental and not on-the-spot.[33]

Yet, all these are minor concerns as compared to the fact that it remains unclear what exactly makes White's notion of the historical event qualify as "historical" in the first place. To understand the difficulties involved, consider that White does not really answer—and does not even attempt to answer—his own question. Remember, the question goes like this: "Can we imagine a new *kind* of event breaking in on our own world which might manifest evidence of another, alternative system of existence that differs utterly from our own?"[34] As excellent as this question may be, White's discussion of "traumatic," "modernist," and "holocaustal" events has nothing to say about how such events manifest "a system of existence that differs utterly from our own," that is, how they imply change over time.

It seems to me that White cannot say anything about change as linked to the notion of the historical event simply because the modernist traumatic event, his theoretical invention, is not really "historical." The traumatic events of the past cause one to be stuck with the past instead of moving forward. For someone being stuck with the traumatic past, the event does not manifest an *alternative* system of existence; instead, it *is* one's very system of existence. The traumatic past may indeed differ "utterly from our own," but this is not because something new happens to those stuck with the past—except at the very moment of the occurrence of the traumatic event.

Let me explain. To begin with the on-the-spot situation, the experienced subject cannot invest the event with meaning and cannot configure it in relation to other events. The traumatic is insular. And the same is true when the event becomes past. As soon as the traumatic event ends, the only way in which the traumatic event differs "utterly from our own" system of existence is that some members of our community are stuck with their past while the world has changed for the rest of the community since that event. But for the experiencing subject it remains an insular event. Instead of being connected to a societal experience of time as historical and changing, a traumatic event belongs to an insular past surviving in the present.

Focusing on the representation of events in their insularity only leads White even further from the question of change. Trying to find adequate means for representing events in their insularity is an effort directed at finding adequate means for rendering insular occurrences sensible without actually making them "historical" in the sense of linking them with other events as part of a larger configuration of change. The only sense in which such insular traumatic events are assumed to be "historical" is simply by virtue of the fact that, as seen from the present, they have taken place in the past.

But such a naïve understanding of being historical is surely not something that White would consciously advocate. The whole tradition of postwar historical theory speaks against reducing "history" to the "past," and we have seen earlier that historical thinking throughout Western modernity has meant seeing individual occurrences, events, and phenomena within a larger trajectory of continuity in change. White's own words can testify that he is very well aware of this:

> Part of the fascination with anything apprehended as "historical" is its appearance of its continuity in change and its change in continuity. Indeed, historical entities have conventionally been conceptualized as undergoing not only metamorphoses but also transubstantiations. Things historical do not "hold steady," do not remain "fixed" so that we can capture them, as it were, in a snapshot or painted portrait.[35]

No notion of the historical event can do without a transformative potential. Whereas Sewell's theory plays out the transformative aspect the most, lacking the attribution of such potential to "modernist" and "traumatic" event questions the extent to which White's events are "historical." The transformation incited by the event does not necessarily need to take the shape of a developmental process as in the modern Western idea of history. It does not need to mean "continuity in change" and "change in continuity." But, in one way or another, as both Gruner and Sewell show, transformation must occur as the result of a specifically "historical" event.

Notes

1. Fernand Braudel, *On History*, trans. Sarah Matthews (Chicago: University of Chicago Press, 1980), 27.
2. For a review of the historiographical debates of the second half of the last century on the theme of events versus structures see Sarah Maza, *Thinking About History* (Chicago: University of Chicago Press, 2017), 157–178. For the closely related phenomenon often referred to as "the revival of narrative" see Lawrence Stone, "The Revival of Narrative: Reflections of a New Old History," *Past & Present* 85, no. 1 (1979): 3–24. Peter Burke, "History of Events and the Revival of Narrative," in *New Perspectives on Historical Writing*, ed. Peter Burke (Cambridge, UK: Polity Press, 1991), 233–248. For an argument about an even more recent "return of the event" see Marek Tamm, "Introduction: Afterlife of Events: Perspectives on Mnemohistory," in *Afterlife of Events: Perspectives on Mnemohistory*, ed. Marek Tamm (Basingstoke: Palgrave, 2015), 6–9.
3. To provide a brief sample, for analytical philosophies of histories see W. B. Gallie, *Philosophy and Historical Understanding* (London: Chatto and Windus, 1964); W. H. Dray, "On the Nature and Role of Narrative in Historiography," *History and Theory* 10, no. 2 (1971): 153–171; Arthur

C. Danto, *Narration and Knowledge: Including the Integral Text of Analytical Philosophy of History* (New York: Columbia University Press, 1985). For phenomenological approaches see Paul Ricœur, *Time and Narrative*, 3 vols., trans. Kathleen McLaughlin and David Pellauer (Chicago: University of Chicago Press, 1984–1988); David Carr, *Time, Narrative and History* (Bloomington: Indiana University Press, 1986). For White's literary theory-inspired narrative philosophy of history see his essays in Hayden White, *Tropics of Discourse: Essays in Cultural Criticism* (Baltimore: Johns Hopkins University Press, 1978); and Hayden White, *The Content of the Form: Narrative Discourse and Historical Representation* (Baltimore: Johns Hopkins University Press, 1987).
4. Rolf Gruner, "The Notion of an Historical Event, I." *Proceedings of the Aristotelian Society, Supplementary Volumes* 43 (1969): 141–152. The article was part of a "symposium" in which William Walsh commented on Gruner's ideas. I think Gruner deserves the credit of a stand-alone citation, but here is the full reference: Rolf Gruner and W. H. Walsh, "Symposium: The Notion of an Historical Event," *Proceedings of the Aristotelian Society, Supplementary Volumes* 43 (1969): 141–164.
5. Hayden White, "The Modernist Event," in *The Persistence of History: Cinema, Television, and the Modern Event*, ed. Vivian Sobchack (London and New York: Routledge, 1996), 17–38. The second one appears in its last version as the third chapter entitled "The Historical Event" in Hayden White, *The Practical Past* (Evanston: Northwestern University Press, 2014), 41–62.
6. William H. Sewell Jr., "Historical Events as Transformations of Structures: Inventing Revolution as the Bastille," *Theory and Society* 25, no. 6 (1996): 841–881.
7. Gruner, "The Notion of an Historical Event," 141.
8. Gruner, "The Notion of an Historical Event," 145.
9. Gruner, "The Notion of an Historical Event," 145.
10. Gruner, "The Notion of an Historical Event," 148.
11. Gruner, "The Notion of an Historical Event," 149.
12. Sewell, "Historical Events as Transformations of Structures," 877.
13. Sewell, "Historical Events as Transformations of Structures," 843.
14. Sewell, "Historical Events as Transformations of Structures," 843.
15. Sewell, "Historical Events as Transformations of Structures," 872.
16. Sewell, "Historical Events as Transformations of Structures," 861.
17. White, *The Practical Past*, 47 (emphasis in the original).
18. White, *The Practical Past*, 47 (emphasis in the original).
19. White, *The Practical Past*, 48 (emphasis in the original).
20. Hayden V. White, "The Burden of History," *History and Theory* 5, no. 2 (1966): 111–134.

21. White, "The Modernist Event," 20.
22. White, "The Modernist Event," 20.
23. White, "The Modernist Event," 24.
24. White, "The Modernist Event," 24.
25. White, "The Modernist Event," 32.
26. White, *The Practical Past*, 57–58.
27. White, "The Modernist Event," 31–32 (emphasis in the original).
28. Rachel G. Hoffman, "The Age of Assassination: Monarchy and Nation in Nineteenth-Century Europe," in *Rewriting German History: New Perspectives on Modern Germany*, eds. Jan Rüger and Nikolaus Wachsmann (Basingstoke: Palgrave, 2015), 121–141.
29. Cf. White's own words: "In order for an account of the events to be considered a historical account, however, it is not enough that they be recorded in the order of their original occurrence. It is the fact that they *can* be recorded otherwise, in an order of narrative, that makes them at once questionable as to their authenticity and susceptible to being considered tokens of reality. In order to qualify as 'historical,' an event must be susceptible to at least two narrations of its occurrence. Unless at least two versions of the same set of events can be imagined, there is no reason for the historian to take upon himself the authority of giving the true account of what really happened. The authority of the historical narrative is the authority of reality itself; the historical account endows this reality with form and thereby makes it desirable, imposing upon its processes the formal coherency that only stories possess." Hayden White, "The Value of Narrativity in the Representation of Reality," *Critical Inquiry* 7, no. 1 (1980): 23.
30. To avoid misunderstandings, I do think that the way Western societies perceive certain events and the way they expect change to take place in certain domains of life defy storytelling and narration. See Zoltán Boldizsár Simon, "The Limits of Anthropocene Narratives," *European Journal of Social Theory*, online first article (2018), https://doi.org/10.1177/1368431018799256. The problem is that White says nothing about how and why exactly a collapse of narrative would happen, and when he says something, he mistakes this claim for the claim that many stories can be told about the one and the same event.
31. Cf. Henry Rousso, *The Latest Catastrophe: History, the Present, the Contemporary*, trans. Jane Marie Todd (Chicago: University of Chicago Press, 2016).
32. For an introduction to ego-histories in English see Jeremy D. Popkin, "*Ego-histoire* and Beyond: Contemporary French Historian Autobiographers," *French Historical Studies* 19, no. 4 (1996): 1139–1167. For a recent collection see *Ego-histories of France and the Second World War*, eds. Manuel Bragança and Fransiska Louwagie (Cham: Palgrave, 2018).

33. To my knowledge, hardly any historian conflates live broadcasts and reportages with their profession. Even if media theory coined the term "media event" to interpret events such as the Olympic Games or presidential debates—and even not pre-staged events such as White's example of the *Challenger* catastrophe—as "the live broadcasting of history," the notion of history here refers to an event that likely will be remembered, not independent of the fact that the event in question is broadcasted live. While those "media events" may become "historical" events, the exact way in which they may do so is a question that nevertheless will be answered after the fact. See Daniel Dayan and Elihu Katz, *Media Events: The Live Broadcasting of History* (Cambridge, MA: Harvard University Press, 1992). On the other hand, while the live broadcasting of "media events" is definitely a new phenomenon, it recently began to capture the attention of historians. In this context, see Espen Ytreberg, "Towards a Historical Understanding of the Media Event," *Media, Culture & Society* 39, no. 3 (2017): 309–324.
34. White, *The Practical Past*, 47 (emphasis in the original).
35. White, *The Practical Past*, 68.

Bibliography

Bragança, Manuel and Fransiska Louwagie, eds. *Ego-histories of France and the Second World War*. Cham: Palgrave, 2018.

Braudel, Fernand. *On History*. Translated by Sarah Matthews, Chicago: University of Chicago Press, 1980.

Burke, Peter. "History of Events and the Revival of Narrative," In *New Perspectives on Historical Writing*, edited by Peter Burke, 233–248. Cambridge, UK: Polity Press, 1991.

Carr, David. *Time, Narrative and History*. Bloomington: Indiana University Press, 1986.

Danto, Arthur C. *Narration and Knowledge: Including the Integral Text of Analytical Philosophy of History*. New York: Columbia University Press, 1985.

Dayan, Daniel, and Elihu Katz. *Media Events: The Live Broadcasting of History*, Cambridge, MA: Harvard University Press, 1992.

Dray, W. H. "On the Nature and Role of Narrative in Historiography." *History and Theory* 10, no. 2 (1971): 153–171.

Gallie, W. B. *Philosophy and Historical Understanding*. London: Chatto and Windus, 1964.

Gruner, Rolf. "The Notion of an Historical Event, I." *Proceedings of the Aristotelian Society, Supplementary Volumes* 43 (1969): 141–152.

Gruner, Rolf, and W. H. Walsh. "Symposium: The Notion of an Historical Event." *Proceedings of the Aristotelian Society, Supplementary Volumes* 43 (1969): 141–164.

Hoffman, Rachel G. "The Age of Assassination: Monarchy and Nation in Nineteenth-Century Europe." In *Rewriting German History: New Perspectives on Modern Germany*, edited by Jan Rüger and Nikolaus Wachsmann, 121–141. Basingstoke: Palgrave, 2015.

Maza, Sarah. *Thinking About History*. Chicago: University of Chicago Press, 2017.

Popkin, Jeremy D. "*Ego-histoire* and Beyond: Contemporary French Historian-Autobiographers." *French Historical Studies* 19, no. 4 (1996): 1139–1167.

Ricœur, Paul. *Time and Narrative*, 3 vols. Translated by Kathleen McLaughlin and David Pellauer. Chicago: University of Chicago Press, 1984–1988.

Rousso, Henry. *The Latest Catastrophe: History, the Present, the Contemporary*. Translated by Jane Marie Todd, Chicago: University of Chicago Press, 2016.

Sewell Jr., William H. "Historical Events as Transformations of Structures: Inventing Revolution as the Bastille." *Theory and Society* 25, no. 6 (1996): 841–881.

Simon, Zoltán Boldizsár. "The Limits of Anthropocene Narratives," *European Journal of Social Theory*, online first article (2018), https://doi.org/10.1177/1368431018799256

Stone, Lawrence. "The Revival of Narrative: Reflections of a New Old History." *Past & Present* 85, no 1 (1979): 3–24.

Tamm, Marek. "Introduction: Afterlife of Events: Perspectives on Mnemohistory." In *Afterlife of Events: Perspectives on Mnemohistory*, edited by Marek Tamm, 1–23. Basingstoke: Palgrave, 2015.

Ytreberg, Espen. "Towards a Historical Understanding of the Media Event," *Media, Culture & Society* 39, no 3 (2017): 309–324.

White, Hayden V. "The Burden of History," *History and Theory* 5, no 2 (1966): 111–134.

White, Hayden. *Tropics of Discourse: Essays in Cultural Criticism*. Baltimore: Johns Hopkins University Press, 1978.

White, Hayden. "The Value of Narrativity in the Representation of Reality." *Critical Inquiry* 7, no. 1 (1980): 5–27.

White, Hayden. *The Content of the Form: Narrative Discourse and Historical Representation*. Baltimore: Johns Hopkins University Press, 1987.

White, Hayden. "The Modernist Event." In *The Persistence of History: Cinema, Television, and the Modern Event*, edited by Vivian Sobchack, 17–38. London and New York: Routledge, 1996.

White, Hayden 2014. *The Practical Past*. Evanston: Northwestern University Press.

CHAPTER 6

The Epochal Event

Abstract In pulling together the threads of this book, this chapter offers a conceptualization of the epochal event as measured against existing notions of the historical event. The chapter begins with a fourfold categorization of events with respect to their historicity: ahistorical events, continuous historical events, discontinuous historical events, and epochal events. In the course of elaborating on the conceptual outlook of the epochal events, the following aspects receive special attention: their hyper-historical character meaning that the transformations they induce opens up a "new reality"; their potential to thereby introduce a radical temporal break interpreted in terms of temporal incommensurability; the way in which they exceed the confines of human experience; and their general extremity. The chapter concludes with a definition of the epochal event.

Keywords Epochal event • Connective concept • Hyper-historical • New reality • Temporal incommensurability • *Sattelzeit* • Human experience

Difference in Kind

Like historical events, the instances of the new epochal introduced in Chapter 4—the potential transgression of planetary boundaries, the prospect of a technological singularity, and the sixth extinction—are vested

Z. B. Simon, *The Epochal Event*, Palgrave Studies in the History of Science and Technology, https://doi.org/10.1007/978-3-030-47805-6_6

with a transformative potential. But the transformations they imply seem far larger than what is commonly associated with historical events and historical change. The changes kicked off by the new epochal are overwhelming and momentous to an extent that was inconceivable before the first nuclear tests, which mark the rise of an awareness of the systemic and profound entanglement of the world of human affairs and the natural world through the unparalleled developments in technology and science.

Differently put, the transformative events of the new epochal signal anthropogenic transformations on a planetary scale in a more-than-human world. The emergence of such events confronts us with the task of making sense of their overall constitution and points to the necessity of justifying their novelty. For serious doubts may be raised concerning whether there is anything qualitatively new here. It is possible to hold the view that scaling up to the level of planetary transformations amounts to nothing more than extending the familiar ways we think about historical change from a human to a more-than-human world. On the ground that epochal events share a transformative potential with historical events, one may argue that the best way to look at epochal events is to view them as larger-scale historical events. Does this sound plausible? Or is there something that clearly distinguishes between epochal and historical events?

In what follows, I will measure the new instances of the epochal against existing notions of the historical event and argue that the difference between them is not merely a difference in degree but a difference in kind. I will attempt to give a conceptual shape to the epochal event as an emerging category of historical thought by developing a fourfold categorization of events with respect to their historicity and by elaborating on a set of interrelated themes. The themes include the hyper-historical character of epochal events; their perceived capacity to introduce temporal incommensurability and thereby to mark a new beginning as an entry into a new reality; the way they escape the confines of human experience; and their general extremity. Eventually, all this will lead to a concluding definition of the epochal event.

A Categorization of Events

Let me begin by sketching a contrast between old and new, between what the work of the previous chapters revealed about historical events and the new epochal, respectively. Sewell's investigation has been the most instrumental in indicating the transformative potential of historical events by pointing out that they introduce "new conceptions of what really exists."[1] The taking of the Bastille was a perfect example inasmuch as in Sewell's interpretation it brought about a new political concept of revolution that articulated popular violence in a legitimate form, as the nation's sovereign will. Contrary to this, what Chapter 4 revealed about the new epochal was that it brings about something much bigger than a new conception of what exists. In the words of Vernor Vinge, describing the notion of technological singularity, the epochal change kicked off by the event results in an entire "new reality" that overwrites all previous models.[2]

Whereas the historical event introduces novelty in the very same perceived reality in which the event itself takes place, the epochal event is conceived of as that which separates two realities. In other words, by marking an entry into a "new reality," the epochal event leads to a completely new world and thereby brings about temporal incommensurability. The post-evental reality is perceived as beings so radically disparate to the pre-evental reality that the two seem to lack a common measure.

The epochal event's perceived capacity to catalyze such a radical rupture clearly departs from Gruner's understanding of the historical event with respect to the necessity of a self-identical subject of change. Gruner made the rather strong claim that "if an object changes its identity it is no longer the same object, and if it is no longer the same object then we involve ourselves in a self-contradiction, i.e., the word change becomes inapplicable."[3] As to the question of the "historical" character of change, Breisach and White expressed a related idea by pointing out that whatever we call "historical" involves both continuity and change.[4] The self-identity of that which goes through changes accounts for the element of continuity, while change takes place in the condition of the self-identical subject.

In the epochal event, however, what is at stake is precisely this self-identity. The epochal event proves Gruner wrong in that it brings about a change that marks the end of previously existent objects. It also questions the assumptions of Breisach and White that historical change over time necessitates continuity. This has indeed been true of the modern Western idea of history, which counterbalanced with a deep underlying continuity even its most "radical" changes. Taking the question of historical transformations out of the framework of continuity may precisely be the novelty of epochal events. Yet, Sewell's notion of the historical event as that which results in something previously inexistent already challenged Gruner's idea and questioned continuity, at least to a certain extent. But the epochal event challenges even Sewell's notion of the historical event in that it brings about a complete separation of two worlds, as seen above. For entering a "new reality" that is inaccessible to pre-evental thought is nothing other than an entry into a world in which the subjects and objects of the new world are not merely new conditions of old subjects and objects and not merely new human conceptualizations; they are new subjects and objects populating a new world.

Finally, there is also White's notion of the historical event as a twentieth-century "modernist" event with a "traumatic" character,[5] which turned out to be ahistorical in the investigations of Chapter 5. The ahistoricity of White's "modernist" event is nevertheless instrumental in fleshing out a fourfold categorization of events with respect to their very historicity. Taking into account all reviewed notions of the historical event, as well as the differences between these notions and the instances of the new epochal, I would like to distinguish between the following categories of events in relation to the question of historicity and change over time: ahistorical events; continuous historical events; discontinuous historical events; and epochal events. The adjectives attached to the categories do not express an inherent quality of the events themselves. Rather, they reflect modalities by which we make sense of the transformations we think such events bring about (or their lack thereof). Let me introduce the categories briefly, one by one.

1. *Ahistorical events.* The first category of ahistorical events consists of events which may be traumatic and thus modernist in the Whitean sense. They do not introduce change and thus are not historical inasmuch as they appear to be insular in time and demand representation without integrating them into a historical trajectory, into a larger scheme or configuration of change over time.
2. *Continuous historical events.* In the second category of continuous historical events, there are the historical events as theorized by Gruner: historical events we conceive of as ones that bring about change in the condition of self-identical subjects. These are the events that feature in most narrative histories written in the last two centuries. For instance, in histories of war, the battles function as such continuous events. And this is true of even those battles that introduce a "turning point" in the course of war. The events associated with turning points make sense precisely on the assumption that there is a course of affairs of deep continuity within which tides turn. And even if these events turn the tides, they do not lead beyond the order of existing things. They turn the tides within the confines of the existing order.

3. *Discontinuous historical events.* Events in the third category do not merely change that which already exists as continuous historical events do. Instead, as a counterpart of continuity, they bring discontinuity into the reality we perceive as our known reality by adding previously inexistent elements to it. The historical events of Sewell belong here, those that introduce change in terms of introducing "new conceptions of what really exists." It is the added novelty that appears as discontinuous with anything pre-existing, much in the way of Sewell's example of the new political concept of revolution. Yet, all this happens without casting a shadow of doubt over the ways in which we conceive of the fundamental constitution of the world itself.
4. *Epochal events.* Events of the second and third category are historical events as we typically look at them. They differ in the way they configure a step from one stage to another, but they do not call into question the logic of gradual developments. Arguably, events are far more frequently evoked as belonging to the second category than to the third. Neither of the categories, however, captures the way in which certain events defy the entire continuity-discontinuity framework by implying a sea change in the way we think about what constitutes our reality in the first place. The events of the fourth category are events like this. The changes they introduce are not gradual changes in a development (either continuous or discontinuous), but entries into incommensurable new realities. They are the ones that I call *epochal events.* By virtue of bringing about a "new reality" of new subjects and objects, they are best labeled as *hyper-historical.*

By labeling epochal events as hyper-historical I do not mean that they are simply more intense and energetic historical events. Nor do I mean that epochal events are hyper-historical because of their planetary scale and more-than-human scope. The epochal event is not merely a bigger historical event, but another *kind* of event, a category on its own. Attributing a hyper-historical character to the epochal means precisely that the changes they bring about are not merely "historical" changes as we have known them throughout Western modernity. The kind of change introduced by epochal events is what I referred to earlier as unprecedented change, that is, a change that does not merely unfold from past conditions but brings about a previously inexistent subject in the shape of a singular event.[6] Aligning the two concepts hints at a concise definition of the epochal event as that which results in unprecedented change. Bringing the two notions together, however, also qualifies my previous use of the notion of unprecedented change in two respects.

First, certain tenets of my previous characterization of unprecedented change as the birth of something previously inexistent were, in principle, far too close to discontinuous historical events as defined above with the help of Sewell's notion. In this respect, linking unprecedented change with epochal events makes clear the extent to which unprecedented change lies beyond the reach of discontinuous historical events. Second, my previous work typically discussed unprecedented change in relation to a new historical sensibility emerging in the ecological and technological domains. Linking the notion with the epochal event makes clear that the proper context to discuss the emergence of unprecedented change as a new historical sensibility is more the human-technology-nature entanglement and less certain domains in themselves.

That said, my purpose here is not to reiterate earlier work but to introduce the epochal event as a new category of historical thought and as a potential connective concept of an emerging knowledge regime designed to study the entangled world of human affairs, technology, and the natural world. To achieve this, there is still a long way to go. The above characterization serves only as a point of departure toward gaining a better understanding of what exactly it means to launch a "new reality" through an epochal event.

Temporal Incommensurability and New Reality

In order to qualify as a connective concept gesturing toward a transdisciplinary knowledge regime, the notion of the epochal event must somehow resonate with already existing disciplines and knowledge formations. Chapter 2 identified the appeal to historical thinking implied by the new epochal as having such a transdisciplinary reach. At the same time, it became equally clear that the appeal to historical thinking entails a demand of renewal: the appeal to history by a new epochality is an appeal to a reconfigured historical thinking that no longer conceives of change over time in terms of developmental processes in the separated human and natural worlds. Finding a resonance with already existing knowledge formations may concern all disciplines invested in historical thinking only inasmuch as they are open to consider alternative scenarios of noncontinuous transformations over time in a more-than-human world.

The difficulty lies with the seemingly contradictory aims of hinting at a *wholesale renewal* and resonating with *existing* ways of knowledge production. But perhaps the contradiction is only apparent. For it must be clear that the epochal event as a connective concept cannot only be linked to a yet-to-be-theorized kind of historical thinking that may explicitly be engaged with issues stretching beyond the confines of the world of human affairs. If the case was so, only the avant-garde few would be able to connect to it in the humanities presently. To be effective as a connective concept, in one way or another, it must make sense even for histories of the human world. Thus, the question is whether it is possible to think of the epochal event as an emerging category of a renewed historical thought in the sense that its scope covers change *both* in the human and in the more-than-human worlds, resonating with both old and new knowledge formations. And the answer I wish to give is not simply that this would be possible, but that—largely without knowing it—this is something we are already doing.

In the post-Second World War intellectual climate, in a tendency simultaneous to the emergence of a new kind of epochal thinking attuned to questions of an entangled human-technological-natural world, the history of the human world has also become seen in terms of large-scale disconnective changes. The two arguably most famous schemes to account for such transformations in the human world concern histories of knowledge in the broadest terms: Thomas Kuhn's *The Structure of Scientific Revolutions* and Michel Foucault's *The Order of Things*, both first published—in English and French, respectively—in the 1960s. Differences aside, they entail a philosophy of history of non-continuous transformations concerning the ways in which societies (by which they both mean Western societies) organize knowledge and conceive of the world.[7]

Comparatively speaking, Foucault's endeavor is the broader one. Whereas Kuhn questions the view that scientific knowledge accumulates over time and argues for an alternative view that looks at the history of science in terms of seldom occurring paradigm changes, Foucault intends to capture the *episteme* of entire periods. The Foucauldian enterprise attempts to apprehend whole knowledge regimes (regulating the conditions of possibilities of what counts as knowledge) as they supersede each other over time. To get a firmer grip on what the succession of *epistemes* looks like, allow me to extensively quote Foucault on his own undertaking:

> I am not concerned, therefore, to describe the progress of knowledge towards an objectivity in which today's science can finally be recognized; what I am attempting to bring to light is the epistemological field, the *episteme* in which knowledge, envisaged apart from all criteria having reference to its rational value or to its objective forms, grounds its positivity and thereby manifests a history which is not that of its growing perfection, but rather that of its conditions of possibility; in this account, what should appear are those configurations within the *space* of knowledge which have given rise to the diverse forms of empirical science. Such an enterprise is not so much a history, in the traditional meaning of that word, as an "archaeology."
>
> Now, this archeological inquiry has revealed two great discontinuities in the *episteme* of Western culture: the first inaugurates the Classical age (roughly half-way through the seventeenth century) and the second, at the beginning of the nineteenth century, marks the beginning of the modern age. The order on the basis of which we think today does not have the same mode of being as that of the Classical thinkers. Despite the impression we may have of an almost uninterrupted development of the European *ratio* from the Renaissance to our own day, despite our possible belief that the classifications of Linnaeus, modified to a greater or lesser degree, can still lay claim to some sort of validity, that Condillac's theory of value can be recognized to some extent in nineteenth-century marginalism, that Keynes was well aware of the affinities between his own analyses and those of Cantillon, that the language of *general grammar* (as exemplified in the authors of Port-Royal or in Bauzée) is not so very far removed from our own – all this quasi-continuity on the level of ideas and themes is doubtless only a surface appearance; on the archeological level, we see that the system of positivities was transformed in a wholesale fashion at the end of the eighteenth and beginning of the nineteenth century.[8]

The all-inclusive transformations in the *episteme* that result in periods in the history of the human world disconnected from each other is, I think, an instance in which a new kind of humanities imagination parallels the equally new kind of epochal thinking concerning the entangled human-technological-natural world in the post-Second World War era.

Just like the new instances of epochal thinking discussed earlier—the ongoing sixth extinction, the potential transgression of planetary boundaries, and the prospect of a technological singularity—Foucault's archeology paints a picture of epochal changes that result in wholly new realities. The new reality, again, just like in the new epochal, is not perceived as that which emerges out of preceding conditions. To Foucault, giving in to the urge of tracing continuity between the new reality and the past means being committed to the "surface appearance." Whereas the succession of historical periods as we came to know them in Western modernity typically entails a common developmental plot (in which, despite all profound changes, the deep continuity of a developmental scenario between the middle ages and modernity is provided by transitional periods such as the renaissance or early modernity), Foucault's periods are successive without having anything like a plot. They come after each other without any substantial connection between them, representing incompatible "systems of possibility," as Ian Hacking puts it in interpreting the theories of both Foucault and Kuhn.[9]

Kuhn's systems of possibility are scientific paradigms. Even though this entails narrower confines than Foucault's *epistemes*, Kuhn has more to say about the relation between the two systems than Foucault. Or, to be more precise, what Kuhn has a term for is the lack of a meaningful relation between scientific paradigms as systems of possibility. The term is *incommensurability*, and Kuhn highlights the incommensurability of scientific paradigms by claiming that "proponents of competing paradigms practice their trades in different worlds": they "see different things when they look from the same point in the same direction."[10] Although paradigms can simultaneously exist and compete, what interests Kuhn the most is their temporal succession, that is, the way in which dominant paradigms are challenged and eventually get superseded by new paradigms. Paradigm shifts are revolutionary periods separating two periods of "normal science," with the latter meaning periods dominated by a ruling paradigm that commands scientific work, which, in turn, is typically devoted to solving puzzles within the frame of the ruling paradigm.

Kuhn's paradigm changes, just as Foucault's shifts in the *episteme*, reflect a mode of thinking that appears to me as the human-world equivalent of the emerging epochal thinking in a more-than-human world. Inasmuch as incommensurability concerns temporally separated paradigms that preside over respective periods of "normal science," paradigm shifts separate two worlds and two realities the same way as the technological singularity or the transgression of planetary boundaries do. In both cases, the post-evental condition is unfathomable to pre-evental thought, and the transformation is anything but a smooth transition.

This is not to say that Kuhn and Foucault deliberately theorized a new mode of epochal thinking within humanities scholarship as applied to changes in the human world with the explicit intention of matching the growing postwar recognition of human capacities bringing about epochal changes on a planetary scale. My claim is only that it is plausible to interpret the theories of Kuhn and Foucault as humanities examples of a broad post-Second World War intellectual tendency to think about large-scale transformations in terms of a previously inconceivable kind of epochal thinking. Inasmuch as the type of transformations is structurally similar, it is even possible to transfer Kuhn's invocation of incommensurability into the world of the human-technology-nature entanglement. The notion of temporal incommensurability is, I believe, a valuable conceptual tool to describe the relationship between pre-evental and post-evental conditions in epochal transformations concerning both the human and the more-than-human worlds.

Beyond the Limits of Human Experience

Despite their shared focus on large-scale transformations, Kuhn and Foucault have little to say about *how* the immense changes they talk about happen. Although it may sound feasible to claim that their theories represent a novel way of thinking about historical change and to associate them with a postwar mobilization of epochal thinking in the humanities, the fact remains that shifts in the *episteme* and paradigm changes focus on the transformations themselves and do not address the possibility of a new kind of event that could signal such transformations. The closest one can get in this respect is Kuhn's analogy between paradigm shifts and gestalt switches. In attempting to illuminate the way in which new paradigms replace old ones, Kuhn resorts to individual conversion experiences and claims that "the transition between competing paradigms cannot be made a step at a time, forced by logic and neutral experience." Instead, on the level of committing individuals, a paradigm change is more "like a gestalt switch, it must occur all at once (though not necessarily in an instant) or not at all."[11]

Although an epochal event is obviously far larger in scale and scope than a personal conversion experience, it would be a mistake to dismiss Kuhn's remark as insignificant for a more profound exploration of the conceptual outlook of the epochal event. To begin with, it is not hard to see how a conversion experience as a gestalt switch may qualify as a transformative event of great significance in one's individual life course. Phenomenologist László Tengelyi even coined the term "destinal events" to name events that introduce such a "radical turn in life-history."[12]

Disregarding the religious and deterministic connotations of the adjective "destinal," the important aspect to note is that Tengelyi tries to capture a radical event that signals a complete "new beginning in life-history."[13]

Marking a "new beginning" is, I think, a structural feature of several notions of events endowed with a transformative character lately. Tengelyi's phenomenological notion is only part of a wider tendency of theorizing transformative events that equally concern individual conversion experiences and socio-political changes. In the former cases, the "new beginning" means the redefinition of personal identity and individual worldviews. In the latter cases, at stake is a "new beginning" for communities. The best examples here may be recent political theories of the event as advocated by prominent thinkers of the political Left.[14] This is not to say that every possible notion of the event marks a new beginning. One can refer to events in various different contexts. In reality, as the previous chapter showed in discussing White's modernist event, even notions that intend to theorize a specifically "historical" event may—mistakenly—lack a transformative potential attributed to events in the first place. What I wish to point out is that insofar as a notion of the event is endowed with a transformative potential, several current theories converge on the point that the transformation they theorize is one that signals a new beginning.

Now, the question is: what kind of a new beginning can be initiated by an event that can reasonably be called epochal? This is the point where Tengelyi can offer no further insights. For a new beginning in life-history, just like Kuhn's gestalt switch, is confined to the realm of first-hand experience. In both Tengelyi and Kuhn, the event is the individual conversion experience itself, while new instances of the epochal typically seem to entail changes that exceed the confines of human experience—and, in some cases, even the confines of human lifetimes. To illustrate the difference: one can convert to seeing things within the Anthropocene framework due to having a first-hand experience of Bruno Latour's 2013 Gifford lectures,[15] but one cannot experience, say, the sixth mass extinction of species (part of the Anthropocene predicament) the same way. The two events are not happening in the same register.

Epochal events confront us with the boundaries of our experience in a twofold way. First, there is the question of the experiencing subject: who exactly is supposed to experience an epochal event if the event seems to concern humans as a biological species? Second, there is the aspect of timescales: an event such as the ongoing sixth mass extinction simply cannot be a matter of individual experience when the event is measured in geological time. Perhaps not all epochal events confront us with the limits of our experience in both ways. The technological singularity, for example, can be a matter of a very short time period, so its timescale may be more familiar. But there is an extent to which the creation of a greater-than-human intelligence it entails is necessarily a concern for us as a species. Either way, none of this comes as a shock for those having a scientific training. But for non-scientists and scholars of the human world, talking about events that somehow entail an extra-experiential dimension likely appears unintelligible.

In his first confrontation with the Anthropocene, Dipesh Chakrabarty already noted both aspects. One of Chakrabarty's initial puzzlements was that making sense of anthropogenic changes in the Earth system requires us to think in terms of a species, even if we cannot experience ourselves in such terms:

> We humans never experience ourselves as a species. We can only intellectually comprehend or infer the existence of the human species but never experience it as such. There could be no phenomenology of us as a species. Even if we were to emotionally identify with a word like *mankind,* we would not know what being a species is, for, in species history, humans are only an instance of the concept species as indeed would be any other life form. But one never experiences being a concept.[16]

For Chakrabarty, this means that the knowledge we produce in trying to make sense of "collective human pasts and futures" in the Anthropocene works "at the limits of historical understanding."[17] Yet this does not mean that we completely lack an experiential relation to the crisis we associate with the Anthropocene. Chakrabarty only thinks that "we experience specific effects of the crisis but not the whole phenomenon," which poses a serious problem. For even if "we may not experience ourselves as a geological agent," there is no escaping the fact that "we appear to have become one at the level of species." Accordingly, even in the humanities, we need to come to terms with the extra-experiential component of the challenge, because "without that knowledge that defies historical understanding there is no making sense of the current crisis that affects us all."[18]

But this is not the only challenge of the Anthropocene to the "experiencing" subject and to knowledge formations grounded in human experience. Chakrabarty is equally mindful to the question of scale. Relying solely on human experience is compromised not only because the current crisis demands an understanding of humanity as a species which he believes we never experience, but also because of the collision of geological time with the time of human history. There is simply no human experience of changes taking place on the geological timescale—at least not with respect to the "whole phenomenon." As Chakrabarty notes in one of his latest articles, geological time "belongs in part to a class of time that has always been seen (long before geology) as opposed to the sense or scale of temporality of human history."[19] Just like the case of thinking in terms of the species, the geological timescale represents for Chakrabarty a "limit to the time of historicality, as a conceptual-temporal place where 'meaning-making' of human history [...] ceases to work."[20]

As pertinent as Chakrabarty's analysis may be, his overall argument may benefit from two crucial qualifications. The first one concerns the relationship between species thinking and human experience. In this respect, Stacy Alaimo delivers a detailed criticism of what she perceives as abstract tendencies in Chakrabarty's position. I think that Alaimo is right in pointing out that it may be possible to experience ourselves in terms of a species in situations of encountering other species.[21] True enough, this still does not enable a first-hand experience of the "whole phenomenon" of the crisis Chakrabarty refers to as the one that exceeds the limits of human experience. But existing forms of species-experience in encountering nonhumans may be helpful in attempting to make sense of other species-level challenges and situating them with concerns of the human world.[22]

The second qualification concerns Chakrabatry's association of the limits of human experience with the limits of historical understanding. As Chapter 2 showed, the Anthropocene predicament and the entanglement of the human, the technological, and the natural worlds indeed upset disciplinary epistemologies. On a more general level, however, historical thinking as a way of making sense of ourselves and the world fares better. As a mode of thinking, modern history has not been confined to the human world: geology's understanding of gradual changes in Earth history and evolutionary theory in biology have been the natural scientific equivalents of disciplinary histories of the human world in Western modernity. Not being confined to the human world means, by entailment, that historical thinking in Western modernity has not been confined to the potential reach of human experience. Again, this remark does not preclude the difficulties concerning the task of situating that which is a matter of human experience with that which is not; it only renders the task a bit easier in our efforts to renew historical thinking.

All in all (and qualifications aside), Chakrabarty's core message seems to be profound and convincing: to get a grasp on the Anthropocene predicament, we need to transcend ways of knowledge production and meaning-making based on an appeal to human experience. The same applies to other instances of epochal thought, from the ongoing sixth extinction to the prospect of transgressing planetary boundaries. They work in an extra-experiential register. To be clear, it is not necessary for each individual epochal event to exceed the confines of human experience in all possible ways. But there is an extent to which all epochal events do so. In one way or another, they exceed the realm of what can be humanly experienced with respect to either their timescale or their entailed "experiencing" subject. Unlike historical events of the human world, epochal events of the more-than-human world do not belong to the order of primary experience.

Extremity

The extra-experiential character of epochal events leads back to the main question this chapter would like to settle. Kuhn and Foucault already testified that it is possible to think about large-scale changes in the human world in the same way as post-Second World War Western societies came to think about epochal changes in the entangled human-technological-natural world, namely, as changes that open the door to "new realities." Would it be equally possible to theorize an epochal event that signals such an epochal change both in the human and in the more-than-human worlds even though the event itself likely transcends the boundaries of human experience? Can the extra-experiential epochal event be theorized as a category of a new kind of historical thought that equally applies to the human world and beyond?

The short answer is yes. The long answer begins in the second qualification on Chakrabarty's endeavor: defying human experience does not necessarily equal defying historical thinking. Although the extra-experiential register is indeed incompatible with most disciplinary epistemologies that attempted to provide foundations for historical studies in the last two centuries or so, modern historical thinking has been relying on categories and approaches that escape the confines of human experience all along. Not only in terms of a distribution of work among the human and natural sciences in historicizing the human and the natural worlds, respectively, but also in terms of the fundamental categories of historical thought. Periods, economic forces, and many structural entities as theorized by social history all lie beyond what experience could claim as its own.

So far so good. But what about the category of event? It can hardly be claimed that events of a historical character have been among those extra-experiential categories. Notions of the event associated with human history indeed do entail an appeal to human experience. Yet it seems to me that nothing prevents the possibility of conceptualizing a kind of event without such an appeal, an event that exhibits the hyper-historical character mentioned earlier. And the task does not even require hardcore theoretical juggling with ideas. It is perfectly enough to see existing categories from a new angle. What we need is only something like a Kuhnian gestalt switch in which certain slabs of time formerly seen as periods in human history are seen afresh as extreme events (events of a duration typically—although not exclusively—longer than human lifetimes) that bring about changes which result in other slabs of time we keep on calling epochs. The only difficulty would lie in differentiating the slabs of time we call events from those we keep on calling ages, periods, and epochs. But this is not a mindboggling difficulty either. Perhaps such differentiation is not even necessary inasmuch as it is possible to say that both kinds of slabs of time are periods, while those that we regard as the ones in which the changes from one reality to another take place also qualify as epochal events. In fact, we already have the category of transitional periods to distinguish periods of transformation from periods to which we attribute more coherence. We only need to find out the extent to which it is plausible to reinterpret certain transitional periods as epochal events.

If this looks far too abstract, consider the period (in the history of Europe) between the mid-eighteenth and mid-nineteenth centuries that Reinhart Koselleck called the *Sattelzeit* (a saddle period). By this, Koselleck meant a threshold period that marked the entry into modernity through spectacular conceptual changes. A large cluster of interrelated concepts—history, revolution, utopia, and so on—has gained a temporal character and thereby new meanings in the *Sattelzeit*, accompanied by the simultaneous invention of new temporal concepts. The net result of interrelated conceptual changes was a completely new understanding of the world as "historical" (the mundane world of affairs changing over time over the developmental course of history), disconnected to the pre-*Sattelzeit* Christian worldview.[23] Much like in a Foucauldian shift in the *episteme* and in a Kuhnian paradigm change, a new world and a new reality was born (a new reality as constituted through the new set of interrelated concepts), incommensurable to the previous one.

The conceptual transformations clustered in an intense period of change that Koselleck calls *Sattelzeit* is, I think, the best example of an epochal event in the human world, kicking off what we usually refer to as Western modernity. It is probably the closest we can get as a human-world equivalent of epochal events in the entangled human-technological-natural world. I do not of course intend to suggest that Koselleck deliberately theorized the *Sattelzeit* period as an epochal event. Nor do I claim that he conceived of it as a transitional period standing in-between and separating two worlds (it seems more plausible that Koselleck thought that, as a threshold period, it forms the opening part of the larger slab of time called the modern period or modernity). But I definitely wish to claim that we can plausibly look at the saddle period *as* a long-term epochal event with an extra-experiential character.

The *Sattelzeit* as an epochal event and the new epochal events in the more-than-human world share an openness to being looked at in dual terms. Think of, again, the ongoing sixth mass extinction of species or mass extinction events in general. They are conventionally referred to as both events and periods. Although the duration of extinction events is a subject of scientific dispute, all are measured in geological time. Compared to events of the human world, they are extremely long. Estimates of the duration of the fifth extinction (the one wiping out dinosaurs) falling within the range of 10,000 years already imply a rapid event on this geological timescale. As to the ongoing sixth extinction, its duration can hardly even be discussed—let alone, settled—in an informed manner precisely because it is ongoing. Even ideas about its beginning can only be blind guesses to a certain extent, because whatever we consider as end points of events have substantial effects on debating their beginnings. But however short the shortest estimates of the duration of the sixth extinction may be, they necessarily remain extremely long as compared to human-world events or even to periods in human history.

The fact of being prone to qualify both as large-scale periods and as events provides a glimpse into the intricacies surrounding the question of the duration of epochal events. Yet the difficulties are not necessarily connected to the length of the event. The technological singularity, the anticipated point at which greater-than-human general intelligence is to be created, is one of the most powerful instances of the new epochal, although its duration is expected to be quite the opposite of a large-scale period.

According to Vinge, it "will be a throwing away of all the previous rules," which can happen extremely quickly, "perhaps in the blink of an eye."[24] Furthermore, the "intelligence explosion" entailed by the singularity, a runaway cycle of creating greater and greater intelligences, "will probably occur faster than any technical revolution seen so far."[25]

All this leaves us with a broad range of options concerning the possible length of epochal events. As they seem to oscillate between the potential instantaneousness of the singularity and the geological time of the anthropogenic sixth extinction, it would be in vain to include considerations of duration among the defining characteristics of the epochal event. What matters more, and what makes an event epochal in the first place, is its perceived capacity to bring about changes that result in a new reality incommensurable to the pre-evental reality.

The key to understanding epochal events lies in their perceived extremity that concerns both period-like long-term and instantaneous short-term versions. For scholars in the humanities and the social sciences—and especially for historians committed to disciplinary codes—coming to terms with such an extremity may be the most challenging. The extremity of epochal events is simply irreconcilable with well-established tools of historical thought, and it questions the core distinctions of historical and social scientific debates on events and structures as mentioned in the previous chapter. It turns upside down and cuts across the established categories of short-term events and long-term structures and their respective associations with the ephemeral and the transient. The dawn of the epochal event marks the collapse of the rigidity of such categorizations. Even efforts that already tried to tone down the event-structure opposition in historical and social thought—like Sewell's notion of the historical event as a trigger event of structural transformations or Immanuel Wallerstein's "world-historical event" that brings about changes in what Wallerstein calls the world system[26]—cannot cope with the extremity of epochal events. On the bright side, none of this entails a general incompatibility with humanities and social scientific scholarship. It rather points to the necessity of a much-needed update in concepts and categories.

Definition

This is the spirit in which I would like to pull together all threads in a definition of the epochal event that I offer as a final contention of this book. The definition is not meant to be set in stone. Its aim is to open a debate on the epochal event as a category of a renewed historical thought and a connective concept. As the definition attempts to capture most of the key points touched upon in the course of the elaboration of the epochal event, it is relatively long. It goes as follows:

> *The epochal event as (1) an emerging category of a new kind of historical thought is best conceived of as (2) a hyper-historical event that (3) brings about a "new reality" and thereby (4) separates two worlds (5) in its capacity to signal the most momentous transformative changes (6) that extend beyond the limits of human experience (7) both in the world of human affairs and in the more-than-human world of the human-technology-nature entanglement.*

Notes

1. William H. Sewell Jr., "Historical Events as Transformations of Structures: Inventing Revolution as the Bastille," *Theory and Society* 25, no. 6 (1996): 681.
2. Vernor Vinge, "The Coming Technological Singularity: How to Survive in the Post-Human Era," in *Vision-21: Interdisciplinary Science and Engineering in the Era of Cyberspace*, Proceedings of a symposium cosponsored by the NASA Lewis Research Center and the Ohio Aerospace Institute, Westlake, Ohio, March 30–31 (1993), 11–22.
3. Rolf Gruner, "The Notion of an Historical Event, I." *Proceedings of the Aristotelian Society, Supplementary Volumes* 43 (1969): 145.
4. Ernst Breisach, *Historiography: Ancient, Medieval and Modern* (Chicago: University of Chicago Press, 1994), 2–3; Hayden White, *The Practical Past* (Evanston: Northwestern University Press, 2014), 68.
5. Hayden White, "The Modernist Event," in *The Persistence of History: Cinema, Television, and the Modern Event*, ed. Vivian Sobchack (London and New York: Routledge, 1996), 17–38: White, *The Practical Past*, 41–62.
6. Zoltán Boldizsár Simon, *History in Times of Unprecedented Change: A Theory for the twenty-first Century* (London: Bloomsbury, 2019), esp. 1–31.
7. Thomas S. Kuhn, *The Structure of Scientific Revolutions* (Chicago: University of Chicago Press, 1996); Michel Foucault, *The Order of Things: An Archaeology of the Human Sciences* (London and New York: Routledge, 2002).
8. Foucault, *The Order of Things*, xxiii–xxiv (emphasis in the original).
9. Ian Hacking, "Michel Foucault's Immature Science," in Ian Hacking, *Historical Ontology* (Cambridge, MA: Harvard University Press, 2002), 87–98.

10. Kuhn, *The Structure of Scientific Revolutions*, 150.
11. Kuhn, *The Structure of Scientific Revolutions*, 150.
12. László Tengelyi, *The Wild Region in Life-History*, trans. Géza Kállay with the author (Evanston: Northwestern University Press, 2004), 81.
13. Tengelyi, *The Wild Region in Life-History*, 81.
14. For a general introduction to the notion of the event in philosophical debates see Slavoj Žižek, *Event: A Philosophical Journey Through a Concept* (Brooklyn and London: Melville House, 2014). In the political-historical context, see especially Alain Badiou, *The Rebirth of History: Times of Riots and Uprisings*, trans. Gregory Elliott (Verso: London, 2012). For a comparison of Sewell's notion of the historical event and Badiou's notion of the political event see Keith Bassett, "Thinking the Event: Badiou's Philosophy of the Event and the Example of the Paris Commune," *Environment and Planning D: Society and Space* 26, no. 5 (2008): 895–910. For the theological underpinnings of recent theories of the political event as disruptive transformations see Jayne Svenungsson, *Divining History: Prophetism, Messianism and the Development of the Spirit*, trans. Stephen Donovan (New York and Oxford: Berghahn Books, 2016).
15. Published in a reworked version as Bruno Latour, *Facing Gaia: Eight Lectures on the New Climatic Regime*, trans. Catherine Porter (Cambridge: Polity, 2017).
16. Dipesh Chakrabarty, "The Climate of History: Four Theses," *Critical Inquiry* 35, no. 2 (2009): 220.
17. Chakrabarty, "The Climate of History," 221.
18. Chakrabarty, "The Climate of History," 221.
19. Dipesh Chakrabarty, "Anthropocene Time," *History and Theory* 57, no. 1 (2018): 22.
20. Chakrabarty, "Anthropocene Time," 23.
21. Stacy Alaimo, *Exposed: Environmental Politics and Pleasures in Posthuman Times* (Minneapolis: University of Minnesota Press, 2016), 150–151. It is equally important to note that Alaimo's criticism often misconstrues Chakrabarty's position and thus is also open to debate. In my understanding, Chakrabarty struggles with bringing the social, cultural, and political categories of an internally divided human world into a meaningful relation with the abstract and universal thinking implied by the geological scale of Earth system changes. Given this decade-long effort to negotiate a relationship between these modes of thinking, it is rather odd to read a growing body of criticism which associates Chakrabarty with a complete denial of his previous work in postcolonial studies. Alaimo gives in to this tendency by associating Chakrabarty with the position according to which "thinking the human species as geophysical force […] precludes attention to social justice" (152). In fact, what motivates Chakrabarty is precisely the

question of how to affirm both, as he made this clear already in his first engagement with the Anthropocene: "How do we relate to a universal history of life – to universal thought, that is – while retaining what is of obvious value in our postcolonial suspicion? The crisis of climate change calls for thinking simultaneously on both registers, to mix together the immiscible chronologies of capital and species history." Chakarabarty, "The Climate of History," 219–220. In more general terms, there is a disconcerting tendency in recent humanities scholarship. Diverse and thoughtful ideas which do not happen to perfectly align with existing humanities categories and dominant agendas are increasingly reduced to and misrepresented as one-dimensional positions with the assumed deliberate aim (at worst) or unintentional effect (at best) to efface the very humanities categories and agendas with which they do not perfectly align. It would likely be more beneficial for humanities scholarship to address this problem in-house before external critiques make use of it (and misuse it) in arguments within the framework of recent debates on the role and necessity of humanities education.

22. On theories of human-animal kinship see the approaches to a multispecies world mentioned in Chapter 2 alongside Alaimo's book. See Donna J. Haraway, *When Species Meet* (Minneapolis: University of Minnesota Press, 2008); Anna Lowenhaupt Tsing, *The Mushroom at the End of the World: On the Possibility of Life in Capitalist Ruins* (Princeton: Princeton University Press, 2015). See also Ewa Domanska, "Ecological Humanities," *Teksty Drugie*, Special Issue – English Edition (2015), 186–210; Nickie Charles, "'Animals Just Love You as You Are:' Experiencing Kinship across the Species Barrier," *Sociology* 48, no. 4 (2014): 715–730.
23. For the notion of *Sattelzeit* see Koselleck's introduction to the massive endeavor of German historians to map the interrelated conceptual changes: Reinhart Koselleck, "Introduction and Prefaces to the Geschichtliche Grundbegriffe," trans. Michaela Richter, *Contributions to the History of Concepts* 6, no. 1 (2011), 1–37. For the endeavor itself see *Geschichtliche Grundbegriffe: Historisches Lexikon zur politisch-sozialen Sprache in Deutschland*, vols. 1–9, eds. Otto Brunner, Werner Conze and Reinhart Koselleck (Stuttgart: Klett-Cotta, 1972–1997). For Koselleck's work on certain concepts see also Reinhart Koselleck *Futures Past: On the Semantics of Historical Time*, trans. Keith Tribe (New York: Columbia University Press, 2004).
24. Vinge, "The Coming Technological Singularity," 12.
25. Vinge, "The Coming Technological Singularity," 14.
26. Immanuel Wallerstein, "The French Revolution as a World-Historical Event," *Social Research* 56, no. 1 (1989), 33–52. On a side-note, it could be interesting to see whether it would be possible to meaningfully situate

Wallerstein's word-system approach with Earth system science on the basis of a shared systems-thinking. If yes, Wallerstein's world-historical event—for which he takes the French Revolution as a major example—may easily prove to be a relatively close affiliate of the epochal event among the existing categories of the humanities and the social sciences (even it is necessarily confined to work in a different register).

Bibliography

Alaimo, Stacy. *Exposed: Environmental Politics and Pleasures in Posthuman Times.* Minneapolis: University of Minnesota Press, 2016.

Badiou, Alain. *The Rebirth of History: Times of Riots and Uprisings.* Translated by Gregory Elliott. Verso: London, 2012.

Bassett, Keith. "Thinking the Event: Badiou's Philosophy of the Event and the Example of the Paris Commune." *Environment and Planning D: Society and Space* 26, no. 5 (2008): 895–910.

Breisach, Ernst. *Historiography: Ancient, Medieval and Modern.* Chicago: University of Chicago Press, 1994.

Brunner, Otto, Werner Conze, and Reinhart Koselleck, eds. *Geschichtliche Grundbegriffe: Historisches Lexikon zur politisch-sozialen Sprache in Deutschland*, vols. 1–9. Stuttgart: Klett-Cotta, 1972–1997.

Chakrabarty, Dipesh. "The Climate of History: Four Theses." *Critical Inquiry* 35, no. 2 (2009): 197–222.

Chakrabarty, Dipesh. "Anthropocene Time." *History and Theory* 57, no. 1 (2018): 5–32.

Charles, Nickie. "'Animals Just Love You as You Are:' Experiencing Kinship across the Species Barrier." *Sociology* 48, no. 4 (2014): 715–730.

Domanska, Ewa. "Ecological Humanities." *Teksty Drugie*, Special Issue – English Edition (2015), 186–210.

Foucault, Michel. *The Order of Things: An Archaeology of the Human Sciences.* London and New York: Routledge, 2002.

Gruner, Rolf. "The Notion of an Historical Event, I." *Proceedings of the Aristotelian Society, Supplementary Volumes* 43 (1969): 141–152.

Hacking, Ian. "Michel Foucault's Immature Science." In Ian Hacking, *Historical Ontology*, 87–98. Cambridge, MA: Harvard University Press, 2002.

Haraway, Donna J. *When Species Meet.* Minneapolis: University of Minnesota Press, 2008.

Koselleck, Reinhart. *Futures Past: On the Semantics of Historical Time.* Translated by Keith Tribe. New York: Columbia University Press, 2014.

Koselleck, Reinhart. "Introduction and Prefaces to the *Geschichtliche Grundbegriffe.*" translated by Michaela Richter, *Contributions to the History of Concepts* 6, no. 1 (2011): 1–37.

Kuhn, Thomas S. *The Structure of Scientific Revolutions*. Chicago: University of Chicago Press, 1996.

Latour, Bruno. *Facing Gaia: Eight Lectures on the New Climatic Regime*. Translated by Catherine Porter. Cambridge: Polity, 2017.

Sewell Jr., William H. "Historical Events as Transformations of Structures: Inventing Revolution as the Bastille." *Theory and Society* 25, no. 6 (1996): 841–881.

Simon, Zoltán Boldizsár. *History in Times of Unprecedented Change: A Theory for the 21st Century*, London: Bloomsbury, 2019.

Svenungsson, Jayne. *Divining History: Prophetism, Messianism and the Development of the Spirit*. Translated by Stephen Donovan, New York and Oxford: Berghahn Books, 2016.

Tengelyi, László. *The Wild Region in Life-History*. Translated by Géza Kállay with the author. Evanston: Northwestern University Press, 2004.

Tsing, Anna Lowenhaupt. *The Mushroom at the End of the World: On the Possibility of Life in Capitalist Ruins*. Princeton: Princeton University Press, 2015.

Vinge, Vernor. "The Coming Technological Singularity: How to Survive in the Post-Human Era." In *Vision-21: Interdisciplinary Science and Engineering in the Era of Cyberspace*, 11–22. Proceedings of a Symposium Cosponsored by the NASA Lewis Research Center and the Ohio Aerospace Institute, Westlake, Ohio, 30–31 March 1993.

Wallerstein, Immanuel. "The French Revolution as a World-Historical Event." *Social Research* 56, no. 1 (1989): 33–52.

White, Hayden. "The Modernist Event." In *The Persistence of History: Cinema, Television, and the Modern Event*, edited by Vivian Sobchack, 17–38. London and New York: Routledge, 1996.

White, Hayden 2014. *The Practical Past*. Evanston: Northwestern University Press.

Žižek, Slavoj. *Event: A Philosophical Journey Through a Concept*. Brooklyn and London: Melville House, 2014.

CHAPTER 7

Coda: A World of Epochal Transformations

Abstract Making sense of our world of epochal transformations is an ongoing concerted effort of various endeavors. Ideas as diverse as Karen Barad's "agential realism," Ulrich Beck's theory of the "metamorphosis of the world," and the appeal to indigenous knowledges in Déborah Danowski and Eduardo Viveiros de Castro's *The Ends of the World* are equally instrumental in capturing various aspects of the larger picture. Although the respective themes they address and the views they express sound reasonable separately, taken together they entail hardly reconcilable and oftentimes contradictory imperatives with respect to the question of how to inhabit the very world they attempt to understand. Following an overview of conflicting views, this book concludes with the takeaway message that learning to inhabit a world of epochal transformations is not simply a matter of understanding that world and producing knowledge about it in light of one imperative or the other. It is also a matter of learning to navigate carefully among conflicting but equally compelling imperatives that underlie the greatest challenges we face today.

Keywords Epochal transformation • Catastrophe • Big history • Agential realism • Metamorphosis • Indigenous knowledges

Z. B. Simon, *The Epochal Event*, Palgrave Studies in the History of Science and Technology,
https://doi.org/10.1007/978-3-030-47805-6_7

Learning to Inhabit

If twenty-first-century life increasingly revolves around a perception of transformative epochal events, then we need to learn to inhabit this world of epochal transformations. Arguably, we already do so, to a certain extent at least. Whether knowingly or not, an epochal sensibility and the general expectation of a multiplicity of epochal events already frame and structure our lives.

This is not to say of course that we live now in anticipation of the upcoming technological singularity or the impending climate apocalypse, similar to the way medieval Europeans lived in anticipation of the Second Coming and the end times.[1] Nor is it to say that we can now consolidate our belief in modern historical teleologies with a newly "discovered" *telos*.[2] The fact that lives are framed and structured by radical transformations does not entail the conscious orientation of each individual life towards a single and final transformative event. Surely, there must be individuals who live their lives by entertaining beliefs that structure their life-history in such ways. But on a general level there can be no teleology involved in the gargantuan character of the numerous anthropogenic changes we currently witness. They simply do not add up to one single transformative event that individuals could elevate into a life-orienting principle. If anything, at moments when we try to make sense of our current condition, we rather grapple with a complexity provided by the many possible and actual ways in which the world seems to change beyond recognition, abruptly and irreversibly.

Nor does the fact that an epochal sensibility increasingly structures our lives mean that the political domain appropriated the idea that the human and the natural worlds are existentially bound together through technology. Neither policy-making on a large scale nor smaller scale politics seem fully accommodated to the new situation. Achieving political recognition of and accommodation to futures of epochal transformation may very well be the practical outcome that many popular actions, citizen movements, policy recommendations of scientific research, and political interventions of humanities and social scientific scholarship aim at, but the fact remains that the political domain as a whole is slow to affirm even the actuality of anthropogenic climate change, which is only one of the many issues at stake.

Everyday activities are, however, increasingly carried out on life-structuring background assumptions, freshly developing ethical stances and epistemic practices grounded in the perception of facing and already living through epochal transformations. From household tasks through shopping to traveling—from Plastic Free July to reducing carbon footprint by reducing flying time—we have begun to attempt to arrange our daily businesses within a frame of a growing awareness concerning our role in altering planetary conditions. Yet these examples represent only individual and conscious lifestyle choices which apply only to those who can afford to worry about such things as their carbon footprint in the first place. And even if the conscious lifestyle choices of the few are accompanied today by implemented policies such as banning or discouraging the use of plastic bags (which constitute a life-structuring enforced measure on a societal level), learning to inhabit the world of epochal transformations demands something more. It points onward to the necessity of developing answers to larger questions and to the task of gaining a profound understanding of our overall predicament.

Learning to inhabit our new condition begins with efforts aimed at exploring the potential realm of possibilities: the forms that societal action can take in facing anthropogenic planetary-scale existential threats to ourselves and other life-forms, the perils of a governance based on measures of pre-emption, or the fate of politics as we know it (meaning not only democratic politics but politics in all its known forms). While a rapidly increasing scholarly literature is available on these themes, addressing such questions about how to inhabit a new world must correlate with simultaneous efforts to gain an understanding of that which we try to inhabit: the new condition itself, the new predicament of the human-technology-nature entanglement, and its epochal transformations. There is no clear order in how to tackle these questions. It is not that first we need to comprehend the overall constellation and our conditions of possibilities so that we become able to address the more specific questions concerning how to inhabit the new constellation. The case is rather that we are grappling with the general and the specific at the same time, while trying to navigate our ways in the midst of unprecedented complexities.

In doing so, each effort has its particular emphases and focal points, and the focus of this book admittedly lies with the general. Its endeavor of exploring the world of a new epochality joins a series of conceptual efforts that likewise grapple with the expansiveness of the transformations amidst which we find ourselves as soon as we recognize the entanglement of the human, the technological, and the natural worlds. There most certainly is a growing tendency in recent scholarship to attempt to understand the big picture, with a varying degree of convergence between the particular endeavors. This final and quasi concluding chapter aims at situating the approach of this book with a few endeavors within this large pool of oftentimes conflicting views that nevertheless represent a concerted effort to come to terms with the question of understanding and inhabiting a world of epochal transformations. As based on a brief overview, the book ends with the message that learning to inhabit such a world might mean nothing other than learning to navigate carefully among conflicting but equally compelling imperatives.

Making Sense of the Epochal

Let me begin by calling attention to an alternative interpretation that may be helpful in clarifying the conceptual work of the previous chapters. Briefly, I argued that in the typically catastrophic future prospects of an entangled human-technological-natural world (that, in a way, still mean prospects catastrophic to us),[3] we witness the constitution of a new kind of event that I call the epochal event. Contrary to this, in analyzing fictional representations, Eva Horn claims in *The Future of Catastrophe* that imaginaries of a climate disaster represent a "catastrophe without event."[4] In Horn's view, climate disaster scenarios do not have an evental character because climate change is a large-scale phenomenon that lacks tangibility: playing out in the long term and involving several disasters, climate change—the overall phenomenon and not its entailed disasters—is notoriously hard to adequately capture by literary and cinematic means.

Horn's views are perfectly in line with the argument of Chapter 6 concerning the ways in which large-scale scenarios of the new epochal typically extend beyond the realm of human experience. How to explain then the diverging interpretations of the evental character of future catastrophes? It seems to me that what Horn means by "catastrophe without event" is actually a catastrophe that has little to do with existing notions of the historical event. In that respect, Horn and this book are still on the same page. The traditional categories of thinking and existing notions of the event in humanities scholarship do not seem to apply to epochal transformations in a more-than-human world.

It seems to me, however, that delinking recent catastrophic imaginaries from the notion of the event remains committed to the tradition of humanities scholarship in which "event" refers to human-world and human-scale occurrences. Besides, the interpretation of climate catastrophe as a "catastrophe without event" may easily overlook the way in which the natural sciences (and thus even natural histories) do in fact refer to "events." Earth system scientists, for instance, have developed the notion of "tipping points" to indicate "when a small change in forcing triggers a strongly nonlinear response in the internal dynamics of part of the climate system," in which "passing a tipping point is typically viewed as a 'high-impact low-probability' event."[5] Not to mention the extinction events in Earth history discussed earlier. True enough, they work on an extra-experiential scale and they may indeed be difficult to present by literary and cinematic means. But this does not mean that they cannot be conceptualized as a new category of events, or are not already enhanced into a conceptual shape. If we want to take the human-technology-nature entanglement seriously, then the concepts we need are concepts that capture the entanglement, even if such concepts work on an extra-experiential scale (without abandoning connection with the human world).

The question of scale is also central to the second related effort I would like to mention: big history. Working on the largest scale possible, big history may appear the ideal candidate to provide a most comprehensive big-picture view. By virtue of its foundational premise to situate the scientific story about the entire universe with the human story on planet Earth, it must be perfectly equipped to grasp the entanglement. The very own genre conventions of big history, however, may stand in the way of living up to such high expectations. How so?

In a nutshell, each individual big history seeks to organize its respective grand story around a single thread that spans through the entire timeline from the birth of the universe to the point where the human world stands now (and beyond). Fred Spier's big history, for instance, tells a story of increasing complexity from the beginning of time to human culture.[6] But the big history endeavor closest to the central theme of this book is that of Cynthia Stokes Brown, with the intention to tell the story of practically everything through the organizing thread of human impact:

> Some subtle emphasis, of course, has to recur to keep any story from dissolving into a muddle. In this book, that underlying theme is the impact of human activities on the planet, as well as the planet's impact on people. When I combined the story of the planet and the people on it, I found that the actions people have taken to keep our offspring increasing have put the planetary environment and its life-forms in grave jeopardy. Summed up in one phrase, this story portrays the "increase of people" rather than the "ascent of man."[7]

One may immediately note that "people" and their growing "impact" are supposed to play a relatively recent and minor part in a story beginning with the Big Bang. Organizing the big story around the human part necessarily comes out as anthropocentric, especially in light of the recent anti-anthropocentric trend of humanities scholarship. But this is only one potential line of criticism, and I am less interested in criticizing big history than in showing the structural difficulties in coping with the central theme of human impact.

Brown's book undoubtedly raised a question already more than a decade ago that, since then, has rapidly gravitated into the center of academic debates. Being published in 2007, the book took up the theme of "human impact" in a humanities-oriented approach a few years before the notion of the Anthropocene conquered the human and social sciences. However, lacking the support of an adequate vocabulary developed in the Anthropocene debate may be one of the main reasons why the book nevertheless struggles with grasping the current predicament. Although Brown's endeavor indeed ventures into the realm of anthropogenic planetary changes (without phrasing the question as such), there is a gap between talking about the "impact" of humans and the planet on one another as separate entities and addressing the human and the natural worlds as entangled. And this gap is likely one between two scientific paradigms in Thomas Kuhn's sense of the term, that is, a gap between wholesale worldviews, in which adherents to "competing paradigms practice their trades in different worlds."[8]

Earth system science most certainly talks about something other than Brown does. In making the case for the formalization of the Anthropocene epoch with a mid-twentieth-century onset, the Anthropocene Working Group argues in its summary that "the base of the proposed Anthropocene time unit is not defined by the beginning of significant human influence upon Earth." What they try to determine instead is the time when human activities "came to have an effect that was both large and synchronous, and thus leave a clear (chrono-) stratigraphic signal,"[9] which they associate with the time of the first tests of nuclear weaponry. And the "large and synchronous effect" here means an effect on the functioning of the Earth system, through human action considered as becoming a natural force.[10] No wonder Clive Hamilton explicitly claims that the whole Anthropocene discussion in the sciences makes sense only on the premise that the Earth is conceived of as an integrated system.[11] In the view of Hamilton—which is very close to that of this book—the Anthropocene concept is inseparable from the Earth system science worldview. It constitutes a new object of knowledge as brought forth by a new knowledge formation, thereby representing an aforementioned Kuhnian paradigm shift. In viewing and studying the Earth as a system, human activity no longer makes sense as an external "impact" on the planet; it occurs instead as an integral systemic component.[12]

Another good reason why talking about human impact may not be the best way to make sense of our current predicament revolves around the question of what "entanglement" means exactly when it comes to the entangled human, technological, and natural worlds. The more profound the entanglement is the less its elements have an easily identifiable existence on their own that could provide the basis for talking in terms of one externally "impacting" another. There cannot possibly be an answer to the question of entanglement stronger than the one entailed in Karen Barad's work. Although Barad developed her "agential realism" in a context other than the Anthropocene debate, it has a common aim with Anthropocene research inasmuch as both attempt to make sense of the intertwinement of human and nonhuman. Barad's theory aims at providing nothing less than foundations for a new ontology, epistemology, and ethics in which meaning and matter, subject and object, human and nonhuman, and nature and human culture are entangled in the strongest sense. According to Barad, "to be entangled is not simply to be intertwined with another, as in the joining of separate entities, but to lack an independent, self-contained existence," meaning that "individuals do not preexist their interactions" but "emerge through and as part of their entangled intra-relating."[13]

Now, what exactly would the consequence of such a strong understanding of entanglement be for our current predicament? Would it mean that society and nature, the human world and the natural world, lacking a "self-contained existence," become completely indistinguishable, let alone, one and the same? And if yes, would this mean that, lacking such a "self-contained existence," we can no longer act on the impending crisis?

In the internal debate among Marxist approaches, Andreas Malm is especially concerned about these questions. In Malm's view, entangling the human and the natural worlds means considering nature and society to be one and the same both in their substance and in their properties, which prevents us from attributing responsibility for environmental degradation in a differentiated human world. To be able to intervene and save whatever can be saved, Malm thinks that we need to retain the distinction between nature and society at least concerning their properties.[14] Now, I doubt that anyone arguing for a strong sense of entanglement would recognize themselves in

Malm's description as being both a substance and a property monist (phrased in a philosophy of mind vocabulary), including Bruno Latour as the chief target of his criticism.[15] Yet the point I wish to make is not about defending Latour or settling the case. What I want to indicate is simply the central importance of the question of how we understand "entanglement" in regard to the human, the technological, and the natural worlds.

And Malm touches upon something crucial in this sense: the way we understand the entanglement is better to be in harmony with our imperative to act on the crisis in which we think we find ourselves. The same applies to our understanding of anthropocentrism and our commitment to develop non-anthropocentric agendas: as Chapter 4 already indicated, we better admit that a sense of anthropocentrism inescapably informs our intentions to intervene and steer the situation for the better (whatever one thinks is better) through planetary-scale human action. We can likely do a better job at synchronizing our concepts with our imperatives. Or maybe both Barad and Malm are right, despite the fact that they seem to hold opposing views, and maybe what we need is to come to terms with the potentially conflicting imperatives of our current situation. More on this possibility later.

For now, it is time to return to Barad. While the purposes of this final chapter do now allow a profound engagement with Barad's truly breathtaking endeavor of reconfiguring the way we think about reality, its transdiciplinarity must be noted. Through situating quantum physics with late twentieth-century social theory, Barad's approach aims at nothing less than accounting for the objective constitution of the world through "intra-actions" that produce even temporality and spatiality.[16] In other words, "agential realism" does not merely concern our conceptions of the constitution of the world and our perceptions of how the world appears to us; it concerns reality itself. Accordingly, Barad's point of departure is not a perplexity resulting from a sense of disorientation in the midst of a perceived radical transformation of the world, but her training in theoretical physics that she puts into conversation with feminist thought, critical posthumanism, and poststructuralism (among other theories in the humanities and social sciences). Being one of the rare instances of a substantial engagement with both scientific and humanities knowledge formations, Barad's theory is extremely well positioned in offering insights concerning the nature of the human-technology-nature entanglement. As indicated above, this must not mean uncritical acceptance, but it has to be granted that Barad's views on the question of the entanglement of the human and the natural world derive from an entanglement of the knowledge formations that have previously been studying those worlds separately.

At the same time, it also must be noted that Barad's theory has less to offer in trying to understand how transformations occur in the new epochality. With respect to this, the most instructive effort may be the unfinished work of the late Ulrich Beck. It grapples with making sense of our current predicament from the viewpoint of a contention that Beck thinks is shared by all the "many people" today who think that "the world is out of joint and it has gone mad." In Beck's view, "the statement on which most people can agree, beyond all antagonisms and across all continents, is: 'I don't understand the world any more.'"[17]

Beck's endeavor shares the suspicion with this book that the perplexity about our current condition is due to a change concerning the way changes themselves occur to us. To describe a newly emerging type of transformation, Beck coins the notion of "the metamorphosis of the world," referring to "an unprecedented historical form of global change that involves two levels, the macro-level of the world and the micro-level of everyday human life."[18] The specifics of "the metamorphosis of the world" reflect the life-long sociological work of Beck, bringing together his new notion with older ones such as "second modernity," "risk society," "individualization," and a shift from "methodological nationalism" to "methodological cosmopolitanism." Even though an emphasis on these supplementary notions tends to steer Beck's attention away from a profound exploration of the human-technology-nature entanglement that lies behind the emergence of the new epochality, his overall endeavor is instructive and offers ideas that complement those of Barad's.

The "metamorphosis of the world," according to Beck, means "more than, and something different from, an evolutionary path from closed to open; it means epochal change of worldviews."[19] True enough, the very same sentence continues with a qualification concerning the specific content of the metamorphosis, namely, that the epochal change of worldviews is "the refiguration of the national worldview."[20] Even if the content of the "metamorphosis of the world" and metamorphosis as a new type of change seem hardly separable in Beck's theory, I would nevertheless insist on an analytical separation between the two. On the one hand, the sheer realization that a new type of perceived change structures twenty-first-century social life is a most crucial insight that resonates with the most recent scholarship that grapples with the challenges of our times. On the other hand, Beck's ideas about a cosmopolitan transformation of the socio-political domain represent his particular normative take on how to invest the metamorphosis with meaning, pointing onward to the question of how exactly to inhabit our world of new epochality.

In sketching his answer, Beck links metamorphosis with climate change by noting that the latter entails "a coalescence of nature, society and politics."[21] Then, in the next step, he links climate change with a cosmopolitan impulse by claiming that

> the surprising momentum of metamorphosis is that, if you firmly believe that climate change is a fundamental threat to all of humankind and nature, it might bring a cosmopolitan turn into our contemporary life and the world might be changed for the better. This is what I call *emancipatory catastrophism*.[22]

The possibility of a cosmopolitan universalism arising out of catastrophism is, needless to say, only one particular idea in a larger debate.[23] For others, the prospects typically look far bleaker, which leads to the last approach I would like to mention among those with which this book has a close affinity.

In their tellingly entitled book *The Ends of the World*, Déborah Danowski and Eduardo Viveiros de Castro recognize that the Anthropocene may indeed constitute an epoch at least in the geological sense. They think that the most interesting aspect of the Anthropocene concept is nevertheless not its epoch-marking character, but that "it points toward the end of epochality as such, insofar as our species is concerned"—"for it is certain that, although it began with us, it will end without us."[24] Based on this contention, Danowski and Viveiros de Castro venture into reviewing imaginaries of the end of the world in a discussion of an impressively broad scope. They map the Anthropocene debate, touch upon recent cinematic and literary renderings of the end-of-the-world theme, review the philosophical premises of speculative realism, assess technophilic accelerationist theories, and, most importantly, engage with the mythologies and mythocosmologies of indigenous knowledges.

At the heart of Danowski and Viveiros de Castro's overall argument is the idea that there is a lot to learn from Amerindian mythocosmologies in our efforts to come to terms with our current crisis.[25] When framing their position in relation to what they call a "Gaia War" between Humans and Terrans, Danowski and Viveiros de Castro draw on the work and terminology of Bruno Latour and Isabelle Stengers.[26] In doing so, they give a noticeably more militant edge to Latour's original idea of a war in which "the *Humans* living in the epoch of the *Holocene* are in conflict with the *Earthbound* [Terrans in another English translation of Latour's French *Terriens*] of the *Anthropocene*."[27]

Danowski and Viveiros de Castro attempt to qualify on the opposing camps of the "Gaia War." While they leave the camp of Humans being more or less as the equivalent of Latour's "Moderns," they criticize Latour for being indefinite on the question of the composition of the Terran camp.[28] At times they claim in a Latourian spirit that the division "is not only internal to our species" and that "the Gaia War opposes two camps or sides populated by humans and non-humans."[29] Other times they prefer to argue with their human emancipatory agenda and point out that "Terrans cannot but be an 'irremediably minor' people," the "people that is missing" and is yet to come.[30] The former qualification is, I believe, more of a gesture to credit Latour's inspiration. It is the latter that resonates fully with their central agenda, as the association of Amerindians and indigenous knowledges with the Terran camp provides the context for outlining what Danowski and Viveiros de Castro think we can learn from Amerindian mythocosmologies in responding to the challenges of our times.

They begin to outline their response by borrowing from Latour the respective mottos of "plus ultra" for Humans and "plus intra" for Terrans.[31] The former represents the Promethean imperatives of modernity, the Age of Discoveries and European expansion, while the latter intends to "steer clear of any fantasy of 'Promethean mastery' or managerial control over the world understood as humankind's Other."[32] According to Danowski and Viveiros de Castro, "plus intra" must mean today "a technology of slowing down, a diseconomy no longer mesmerized by the hallucination of continuous growth," and even "uncivilization" as a "less titanic project."[33] This is the imperative to which Amerindian and indigenous knowledge can contribute, both through their mythologies and by virtue of the fact that "for the native people of the Americas, the end of the world already happened – five centuries ago," in 1492.[34]

Navigating Among Contradictory Imperatives

Bringing much neglected indigenous knowledges to the fore would likely be welcomed by many in the larger debate. On a critical note, however, it is hard to sidestep the fact that such knowledges are supposed to contribute to the not very indigenous anti-anthropocentric agenda of critical humanities scholarship reviewed earlier. Besides, the contribution itself does not seem specific to indigenous knowledges. Danowski and Viveiros de Castro's project of uncivilization and "technology of slowing down" as pitched against accelerationism and large-scale Promethean technology is in perfect harmony with Eileen Crist's advocacy of "pulling back" and "scaling down" as set against techno-managerial solutions, discussed in Chapter 4.[35]

Pointing at the unspecificity of the role assigned to indigenous knowledges by no means intends to discredit its imperatives. It only sees them within the context of sharing those imperatives with a larger family of approaches and views, resulting in affiliations and similarities that may seem odd and, to a certain extent, even bizarre. And the oddity of alliances with respect to particular ideas and goals becomes even more apparent as soon as Promethean projects enter the picture. For it must be clear that we live in an age of plans of space colonization, asteroid mining, geoengineering, biomedical engineering, technologies to produce cultured meat, and so forth. Whether one likes this or not, is another question.

Promethean projects are unlikely to halt to anytime soon. Anti-Promethean stances are equally unlikely to erode in the near future. What's more, the conflict of imperatives between Promethean and anti-Promethean projects and solutions to anthropogenic crises and scarcities does not mean that the two sides could boil down to an easy formula of opposing camps that disapprove of anything the other side has to offer. The oddity of alliances among the views concerning individual Promethean and anti-Promethean projects rather implies that we occasionally cut across the larger platforms of conflicting imperatives. One can easily have an anti-Promethean stance on space colonization and simultaneously advocate a Promethean view with respect to the medical application of enhancement technologies. The advocates of particular views may push their individual agendas in a focused approach, but in our everyday lives, in different life situations, we typically deal not only with one issue—or, at best, a few issues—as the advocates do. For our lives, all particular issues matter. Anthropogenic changes in the condition of the Earth system, artificial intelligence scenarios, or, for that matter, the singularity concern us just as well as the sixth extinction or genetically modified food production do. However, there is no view and no available knowledge to encompass even a tiny fragment of all issues that have the potential to radically alter our lives. While all issues and all views with their contradictory imperatives concern us, they hardly add up to a coherent picture.

The point I want to make in the shape of a takeaway message is that the question may not be that of whether we should advocate either Promethean or anti-Promethean solutions. Learning to inhabit a world of epochal transformations is not simply a matter of understanding that world and producing knowledge about it in light of one imperative or the other. It is also a matter of learning to navigate carefully among conflicting but equally compelling imperatives that underlie the greatest challenges we face today—the challenges that we struggle to grasp and fathom as one overall challenge.

To see the conflicting imperatives at work, take just one example: the anthropogenic extinction of species, the sixth mass extinction. "Pulling back," "scaling down," and uncivilization as anti-Promethean recommendations to effectively respond to the crisis are counterbalanced by the large-scale and profoundly Promethean project of de-extinction. The latter refers to efforts to reverse extinction by reintroducing species through DNA recovery, cloning, or the genetic engineering of existing species.[36] Controversial as the idea may be, putting scientific knowledge and new genomic technologies into work with the intention to engineer planetary life is just as much on the overall agenda as is the anti-Promethean humanities drive. And, again, I do not see how the Promethean impulse would fade away. In fact, I do not even see how advocates of anti-Promethean agendas would actually be willing to give up on the benefits such enterprises grant on a societal level. And even if I was proven wrong about this, there remains the extent to which giving up on Promethean projects on the large scale demands the workings of the very Promethean powers, as Chapter 4 argued. On the other hand, it is equally clear that the very same enterprises are instrumental in effectively destroying even the benefits they bring about, which is precisely why anti-Promethean measures are demanded.

Neither of the conflicting sides appear to be completely wrong or out of line. At the same time, neither of them seem to offer solutions that could match or even come close to the complexity of the issues to which they try to respond. In this situation, I think that the best we can do is to listen to both sides, to both Promethean and anti-Promethean imperatives as they manifest in a cacophony of approaches and stances.

Given our inability to comprehend the complexity of the larger picture, today may be less of the time of advocating any particular stance against all others and more of the time of trying to navigate among the conflicting imperatives of multiple partially valid and useful stances. This is the only position that I can advocate. The above views, agendas, and conceptual efforts are all indispensable in one way or another, despite their occasional irreconcilability. As the fact that none of them are completely wrong does not mean that all of them are right, they will likely generate heated discussions in a condition of general uncertainty. For now, what may seem a bit more certain is only that learning to inhabit our world of epochal transformations requires us to make the most out of living with contradictory imperatives.

Notes

1. That said, drawing theological and eschatological parallels constitute a major interpretive framework in recent scholarly literature on technological and environmental prospects. See, for instance, Stefan Skrimshire, "Climate Change and Apocalyptic Faith," *Wiley Interdisciplinary Reviews: Climate Change* 5, no. 2 (2014): 233–246; Michael S. Burdett, *Eschatology and the Technological Future* (London and New York: Routledge, 2015). Delf Rothe, "Governing the End Times? Planet Politics and the Secular Eschatology of the Anthropocene," *Millennium: Journal of International Studies* 48, no. 2 (2020): 143–164.
2. On teleology and the modern idea of history see most recently *Historical Teleologies in the Modern World*, eds. Henning Trüper, Dipesh Chakrabarty and Sanjay Subrahmanyam (London: Bloomsbury, 2015).
3. When talking about bleak future prospects in an entangled human, technological, and natural world we typically keep on talking about our future. In the words of Claire Colebrook, "for all the talk of climate change we assume that the climate is what environs us, and that change – or the danger of change – needs to be calculated according to the degree to which it enables or precludes ongoing existence of humans." Claire Colebrook, "Framing the End of Species," *symplokē* 21, no. 1–2 (2013): 59. See also Chapter 4 and the discussion of anthropocentrism even in anti-anthropocentric agendas.
4. Eva Horn, *The Future as Catastrophe: Imagining Disaster in the Modern Age*, trans. Valentine Pakis (New York: Columbia University Press, 2018), 55–88.
5. Timothy M. Lenton, "Early Warning of Climate Tipping Points," *Nature Climate Change* 1 (2011): 201.
6. Fred Spier, *Big History and the Future of Humanity* (Malden, MA: Wiley-Blackwell, 2010).
7. Cynthia Stokes Brown, *Big History: From the Big Bang to the Present* (New York: The New Press, 2007), xii.

8. Thomas S. Kuhn, *The Structure of Scientific Revolutions* (Chicago: University of Chicago Press, 1996), 150.
9. Zalasiewicz et al., "The Working Group on the Anthropocene," 57.
10. For more on the Earth system science understanding of the Anthropocene by members of the Anthropocene Working Group see also Will Steffen et al., "The Anthropocene: Conceptual and Historical Perspectives." *Philosophical Transactions of the Royal Society A* 369 (2011): 842–867; Colin N. Waters et al., "A Stratigraphical Basis for the Anthropocene?" *Geological Society, London, Special Publications* 395, no. 1 (2014), 1–24; Will Steffen et al., "Stratigraphic and Earth System Approaches to Defining the Anthropocene." *Earth's Future* 4 (2016): 324–345; Jan Zalasiewicz et al., "Making the Case for a Formal Anthropocene Epoch: An Analysis of Ongoing Critiques." *Newsletters on Stratigraphy* 50, no. 2 (2017), 205–226.
11. Clive Hamilton, *Defiant Earth: The Fate of Humans in the Anthropocene* (Cambridge: Polity, 2017), esp. 1–35.
12. Cf. Chapter 3 and the discussion of the ways in which human-produced technology appears as a systemic element in scientific Anthropocene-thinking. The same chapter also features a brief discussion on the short-comings of the vocabulary of "human impact" in making sense of our current predicament.
13. Karen Barad, *Meeting the Universe Halfway: Quantum Physics and the Entanglement of Matter and Meaning* (Durham: Duke University Press, 2007).
14. Andreas Malm, "Against Hybridism: Why We Need to Distinguish between Nature and Society, Now More than Ever," *Historical Materialism* 27, no. 2 (2019): 156–187.
15. Malm's critique is focused on Bruno Latour, *We Have Never Been Modern*, trans. Catherine Porter (Cambridge, MA: Harvard University Press, 1993). Aside from occasional mentions, it does not engage with Latour's late work on anthropogenic changes in the Earth system within his preferred Gaia framework, to which I will return later.
16. See especially Barad, *Meeting the Universe Halfway*, 132–185.
17. Ulrich Beck, *The Metamorphosis of the World* (Cambridge: Polity, 2016), xi.
18. Beck, *The Metamorphosis of the World*, 51.
19. Beck, *The Metamorphosis of the World*, 5.
20. Beck, *The Metamorphosis of the World*, 5.
21. Beck, *The Metamorphosis of the World*, 40.
22. Beck, *The Metamorphosis of the World*, 35 (emphasis in the original).
23. To be as clear as possible, by introducing Beck's views I do not wish to suggest that cosmopolitanism is necessarily the kind of social transformation that matches the growing perception of anthropogenic planetary

changes and toward which we are inevitably heading. In a hypothetical scenario in which one did suggest this, it would be reasonable to argue that the recent resurgence of nationalist sentiments in the political landscape of the Global North (clearly pointing to a direction opposite to cosmopolitanism) is, actually, nothing other than a temporary backlash and a counter-reaction to the political constitution of an increasingly globalized world that, eventually, has to take place in the cosmopolitan shape inasmuch as we want to be able to tackle planetary-level challenges. This looks like a decent position. But again, this would be one among the many possible positions, and, in the present political climate, it would gesture toward sounding like wishful thinking. And that which may be desired pragmatically should not be confused with understanding that which is the case. Although normative agendas unavoidably and necessarily pervade the way we conceive of what there is, there must be a fine—even if fluid, hardly identifiable and constantly renegotiated—line between wishful thinking and social analysis. Not to mention that cosmopolitanism is a very old way to think about a universal notion of humanity, inextricably bound to the modern sense of historicity, the unfolding and potential fulfillment of a human potential over the course of a processual historical process, from which even Beck's "metamorphosis of the world" distances itself. On the other hand, there can be new meanings attributed to cosmopolitanism, which is precisely what Beck aims at. So do Gerard Delanty and Aurea Mota in arguing that the Anthropocene predicament points onwards to a new cosmopolitics. See Gerand Delanty and Aurea Mota, "Governing the Anthropocene: Agency, Governance, Knowledge," *European Journal of Social Theory* 20, no. 1 (2017): 9–38.

24. Déborah Danowski and Eduardo Viveiros de Castro, *The Ends of the World*, trans. Rodrigo Nunes (Cambridge: Polity, 2017), 5.
25. See also *Indigenous Perceptions of the End of the World: Creating a Cosmopolitics of Change*, ed. Rosalyn Bold (Cham: Palgrave, 2019).
26. Bruno Latour, *Facing Gaia: Eight Lectures on the New Climatic Regime*, trans. Catherine Porter (Cambridge: Polity, 2017); Isabelle Stengers, *In Catastrophic Times: Resisting the Coming Barbarism*, trans. Andrew Goffey (Open Humanities Press, 2015). Both Latour and Stengers argue within a framework of the Gaia hypothesis, developed by James Lovelock in the 1960s. Today, the Gaia hypothesis is oftentimes credited as the forerunner of the Earth system science view in considering the Earth as one integrated and interactive system. See James Lovelock, *Gaia: A New Look at Life on Earth* (Oxford: Oxford University Press, 2000). Recently, Lovelock even ventures into linking his earlier thoughts with technological prospects of creating greater-than-human intelligence. See James Lovelock, *Novacene: The Coming Age of Hyperintelligence* (Penguin, 2019).

27. Latour, *Facing Gaia*, 248 (emphasis in the original).
28. Danowski and Viveiros de Castro, *The Ends of the World*, 93.
29. Danowski and Viveiros de Castro, *The Ends of the World*, 100.
30. Danowski and Viveiros de Castro, *The Ends of the World*, 94; 123.
31. Latour, *Facing Gaia*, 291.
32. Danowski and Viveiros de Castro, *The Ends of the World*, 97.
33. Danowski and Viveiros de Castro, *The Ends of the World*, 97–98.
34. Danowski and Viveiros de Castro, *The Ends of the World*, 104.
35. Eileen Crist, "Reimagining the Human," *Science* 362, Issue 6420 (2018): 1242–1244; Eileen Crist, *Abundant Earth: Toward and Ecological Civilization* (Chicago: University of Chicago Press, 2019).
36. For a brief introduction of de-extinction see Dolly Jørgensen, "Reintroduction and De-extinction," *BioScience* 63, no. 9 (2013), 719–720; Beth Shapiro, *How to Clone a Mammoth: The Science of De-extinction* (Princeton: Princeton University Press, 2015); Amy Lynn Fletcher, *De-extinction and the Genomics Revolution: Life on Demand* (Cham: Palgrave, 2019).

Bibliography

Barad, Karen. *Meeting the Universe Halfway: Quantum Physics and the Entanglement of Matter and Meaning*. Durham: Duke University Press, 2007.

Beck, Ulrich. *The Metamorphosis of the World*. Cambridge: Polity, 2016.

Bold, Rosalyn, ed. *Indigenous Perceptions of the End of the World: Creating a Cosmopolitics of Change*. Cham: Palgrave, 2019.

Brown, Cynthia Stokes. *Big History: From the Big Bang to the Present*. New York: The New Press, 2007.

Burdett, Michael S. *Eschatology and the Technological Future*. London and New York: Routledge, 2015.

Colebrook, Claire. "Framing the End of Species." *symplokē* 21, no. 1–2 (2013): 51–63.

Crist, Eileen. "Reimagining the Human." *Science* 362, Issue 6420 (2018): 1242–1244.

Crist, Eileen. *Abundant Earth: Toward and Ecological Civilization*, Chicago: University of Chicago Press, 2019.

Danowski, Déborah, and Eduardo Viveiros de Castro. *The Ends of the World*. Translated by Rodrigo Nunes. Cambridge: Polity, 2017.

Delanty, Gerard, and Aurea Mota, "Governing the Anthropocene: Agency, Governance, Knowledge." *European Journal of Social Theory* 20, no. 1 (2017), 9–38.

Fletcher, Amy Lynn. *De-Extinction and the Genomics Revolution: Life on Demand*. Cham: Palgrave, 2019.

Hamilton, Clive. *Defiant Earth: The Fate of Humans in the Anthropocene*. Cambridge: Polity, 2017.

Horn, Eva. *The Future as Catastrophe: Imagining Disaster in the Modern Age*. Translated by Valentine Pakis. New York: Columbia University Press, 2018.

Jørgensen, Dolly. "Reintroduction and De-extinction." *BioScience* 63, no. 9 (2013), 719–720.

Kuhn, Thomas S. *The Structure of Scientific Revolutions*. Chicago: University of Chicago Press, 1996.

Latour, Bruno. *We Have Never Been Modern*. Translated by Catherine Porter. Cambridge, MA: Harvard University Press, 1993.

Latour, Bruno. *Facing Gaia: Eight Lectures on the New Climatic Regime*. Translated by Catherine Porter. Cambridge: Polity, 2017.

Lenton, Timothy M. "Early Warning of Climate Tipping Points." *Nature Climate Change* 1 (2011): 201–209.

Lovelock, James. *Gaia: A New Look at Life on Earth*, Oxford: Oxford University Press, 2000.

Lovelock, James. *Novacene: The Coming Age of Hyperintelligence*. Penguin, 2019.

Malm, Andreas. "Against Hybridism: Why We Need to Distinguish between Nature and Society, Now More than Ever." *Historical Materialism* 27, no. 2 (2019): 156–187.

Rothe, Delf. "Governing the End Times? Planet Politics and the Secular Eschatology of the Anthropocene." *Millennium: Journal of International Studies* 48, no. 2 (2020): 143–164.

Shapiro, Beth. *How to Clone a Mammoth: The Science of De-extinction*. Princeton: Princeton University Press, 2015.

Skrimshire, Stefan. "Climate Change and Apocalyptic Faith," *Wiley Interdisciplinary Reviews: Climate Change* 5, no. 2 (2014): 233–246.

Spier, Fred. *Big History and the Future of Humanity*. Malden, MA: Wiley-Blackwell, 2010.

Steffen, Will et al. "The Anthropocene: Conceptual and Historical Perspectives." *Philosophical Transactions of the Royal Society A* 369 (2011): 842–867.

Steffen Will et al. "Stratigraphic and Earth System Approaches to Defining the Anthropocene." *Earth's Future* 4 (2016): 324–345.

Stengers, Isabelle. *In Catastrophic Times: Resisting the Coming Barbarism*. Translated by Andrew Goffey. Open Humanities Press, 2015.

Trüper, Henning, Dipesh Chakrabarty and Sanjay Subrahmanyam, eds. *Historical Teleologies in the Modern World*. London: Bloomsbury, 2015.

Waters, Colin N. et al. "A Stratigraphical Basis for the Anthropocene?" *Geological Society, London, Special Publications* 395, no. 1 (2014), 1–24.

Zalasiewicz, Jan et al. "The Working Group on the Anthropocene: Summary of Evidence and Iterim Recommendations." *Anthropocene* 19 (2017): 55–60.

Zalasiewicz, Jan et al. "Making the Case for a Formal Anthropocene Epoch: An Analysis of Ongoing Critiques." *Newsletters on Stratigraphy* 50, no. 2 (2017), 205–226.

Index[1]

[1] Note: Page numbers followed by 'n' refer to notes.

Z. B. Simon, *The Epochal Event*, Palgrave Studies in the History of Science and Technology,
https://doi.org/10.1007/978-3-030-47805-6

The manufacturer's authorised representative in the EU is Springer Nature Customer Service Centre GmbH, Europaplatz 3, 69115 Heidelberg, Germany. If you have any concerns regarding our products, please contact ProductSafety@springernature.com

Printed and bound by CPI Group (UK) Ltd, Croydon, CR0 4YY
30/09/2026
02225609-0013